MANAGEMENT: AILMENTS & INJURIES

HORSEKEEPING

MANAGEMENT: AILMENTS & INJURIES

Ray Saunders

Frederick Muller Limited
London

General Editor: Candida Hunt

In the same illustrated series
Horsekeeping 1: Ownership, Stabling & Feeding
Horsekeeping 2: Management: Ailments & Injuries
Horsekeeping 3: Handling & Training
Horsekeeping 4: Small Scale Breeding
Horsekeeping 5: All Aspects in Pictures

First published in UK by
Frederick Muller Ltd
Dataday House
8 Alexandra Road
Wimbledon, London SW19 7JU

Produced for the publisher by Midas Books
12 Dene Way, Speldhurst
Tunbridge Wells, Kent TN3 0NX
England

© Ray Saunders 1982

All rights reserved. No part of this publication may be reproduced, stored in a retrieval system, or transmitted, in any form or by any means, electronic, mechanical, photocopying, recording or otherwise, without the prior permission of Midas Books.

ISBN 0 584 95039 X (Hardcover)
ISBN 0 584 95040 3 (Softcover)

Printed in Great Britain by
Chambers Green Ltd, Tunbridge Wells, Kent.

Contents

		Page
1	Good management practices	9
2	Managing feedstuffs	17
3	Digestion and digestive disturbances	24
4	Wounds and skin disorders	33
5	Coughs and colds	48
6	Whistling and roaring	55
7	Determining lameness	59
8	Troubles affecting the lower leg	66
9	Ailments of the foot	89
	Index	107
	Colour plate section	between pp 48-49

Acknowledgements

My thanks again to Jack Neal for working with me to produce all the photographs for this book and for his endless patience to ensure that each one clearly showed the subject being explained. The result of this is to ensure that the reader obtains the greatest amount of understanding from the text.

My grateful thanks also to the team of Somerset vets, Messrs G. R. Carter, B.V.Sc., M.R.C.V.S., P. C. Browne B.V.Sc., M.R.C.V.S., and W. R. G. West, M.R.C.V.S., who allowed me to photograph operations and injuries as well as providing much helpful advice and information.

Finally I must once more extend my gratitude to my editor, Candida Hunt, for her work in ensuring that my text finished up concise and complete, for which her very considerable knowledge of the subject was of no little importance.

PUBLISHER'S NOTE

This publication is not to be used in place of a veterinary surgeon, but merely to guide you in helping the injured animal until you get professional care. Our guidelines are based on what will happen with most horses and ponies in most situations. However, there are always a few exceptions where the horse or pony may not respond to your help as expected; these animals will need professional care even sooner.

1 Good management practices

Horse and pony management will mean different things to different people, depending on whether the owner has a pleasure animal for hacking and weekend use or, at the other end of the scale, is keeping a competition animal for high-class events. Wherever you come on the scale the desired result is to keep your horse or pony sound, healthy and contented. This is made easier today with modern veterinary techniques and drugs, but I do not believe in using these methods unnecessarily or as a first resort when simpler treatment can first be tried. Expensive treatments can often be avoided with proper care and forethought, and I shall be explaining this fully in later chapters. Apart from the inconvenience to both you and the animal, veterinary services are costly, and prevention is definitely better than cure!

When troubles do occur the correct diagnosis of the problem is vital to determine what treatment is needed and whether the vet need be sent for. This book in no way pretends to be a veterinary journal, nor can it teach anyone to diagnose correctly a condition that may sometimes cause even an experienced vet to be unsure. What I set out to do here is to provide the owner with an understanding of practices and procedures that can lead to trouble, so by avoiding them he can prevent the trouble arising. Also, by giving an explanation in layman's terms of the various workings of the horse's system I hope to assist owners to distinguish between injuries and disorders that can be simply and effectively treated and those requiring more specialized treatment.

The single most important thing in the management of a horse or pony to keep it sound and healthy is forethought. A good rule to adopt is never to do anything without first considering what such action will lead to. This applies equally to feeding and to exercise; and when any changes in routine are contemplated be sure these are carried out gradually. Violent changes of any kind will always be unsettling, and sometimes very harmful, to the horse or pony. Although some people would appear to get away with it and some animals will tolerate abuse better than others, thoughtless practices will nevertheless nearly always be paid for in the end with a sick or injured animal, and usually with a vet's bill as well. Common examples of bad management of which even quite experienced horse owners are sometimes guilty are: sudden changes in the amount or type of feed; greatly increased exercise without proper preparation or with an unfit animal; sudden lessening of exercise with a fit horse without a period of letting down.

Avoiding boredom

Horses and ponies love routine and will be kept healthy and happy on a regular diet and exercise once an individual's specific requirements have been understood. This is not to say that it will not become bored if subjected to the same dreary routine or idleness day after day without change. One's ingenuity must be used within a framework of regular feeding and exercise to keep the animal 'fresh' and to introduce variations that add interest without deviating from the overall feeding and exercise programme. With a stabled horse or pony boredom can present a problem to owners who have to leave their animals penned up for long periods. It is during these times that bad habits such as crib biting, wind sucking and weaving are developed to alleviate boredom. It is sometimes suggested that hanging toys for the animal to play with is helpful, but I am myself not in favour of this as I feel it could teach a horse to bob and weave its head or suck things. In my view it is better to try and break up the animal's day with some kind of attention, and when grooming and exercise are not possible and it must remain stabled try to feed the bulk of its hay ration during the lonely hours to keep it occupied. A radio left on for a few hours might also help to provide 'company', and if not at least it will do no harm. Do try to have the loose-box sited so that the horse or pony can look out and see things going on, and if this includes the activities of other horses or other animals so much the better. I have known the antics of a few ducks on a nearby pond keep a horse contented for part of the day. Front and rear top doors that can be left open, especially in a large box, are also very helpful as the animal can walk around the loose-box and find an alternative view to hold its attention. This is not so good if the doors are sited to admit a blast of cold air across the animal's back on a cold winter's day, but even this is preferable to a dark and stuffy interior with no outlook.

Stabling advantages

In spite of the disadvantages of boredom for the horse or pony and the extra work involved in management for the owner, there are several advantages to be gained by keeping an animal 'in' for much of the time. If you are an owner without the advantage of access to adequate grass and have to keep your animal stabled for much of its life you can draw comfort from this. A properly cared for horse or pony will benefit in a number of ways from being stabled as also will you. For instance, you will not be faced with having to provide winter protection in the form of outdoor rugs. These can cause the animal considerable discomfort if not looked at regularly and properly adjusted, and can on occasions lead to injury. There are inevitably many other things that lead to injury when an animal is turned out, even when precautions have been taken. Ask any horse owner to think of past injuries suffered by their animals and in the majority of cases these will prove to have happened when the animal was turned out. The stabled horse or pony will also keep cleaner; the main cause of mud fever will have been removed in winter; in summer the animal will not be so bothered by flies and will not ingest the eggs that produce bot larvae, and worms will be much less of a problem. The owner without grass will also save time and money on grassland upkeep or hedge and fence management. One need not worry too much, therefore, if grass space cannot be provided so long as one has the time and facilities to keep an animal properly stabled.

Warmth and comfort

The type of bedding used for the stabled horse or pony affects not only its warmth and comfort but also its general health and soundness. The choice of bedding material should be carefully considered in relation to its availability and its suitability to the animal. Straw is the best bedding material in my opinion, and I particularly favour well-managed, deep-littered wheat straw. Just any old straw will not do; dusty straw is undesirable, and that containing mould or spores must never be used as it will prove very injurious to the animal's wind.

Where respiratory problems are present a better bedding medium will be shavings or peat, sometimes the animal's respiratory problem will be such that only keeping it out in the air will help its condition. With any type of bedding adequate ventilation and ample provision of good clean air in the stable is of vital importance. If illness requires the animal to be kept stabled when the weather is wet and windy, a properly ventilated loose-box free from draughts should be provided. A good, deep, resilient bed is a must, and it must be frequently attended to for the removal of droppings and a topping up of fresh, clean bedding. During a stabled convalescence do try to break the animal's boredom by a couple of periods of exercise, even if this is merely confined to short walks around the yard. Don't leave a horse standing for days on end with only the stable wall to look at.

Keeping a sick horse or pony warm in a stable is sometimes done by raising the internal temperature, perhaps by electrical heat lamps, rather than by adding extra clothing, which will be apt to fatigue the animal. Clothing should never be overloaded in the belief that the animal is being kept nice and warm, but one must also be careful not to produce a hot, stuffy atmosphere if heating of any kind is used. Heat lamps used as a treatment for certain types of ailments, especially those involving joints and muscles, can be beneficial, but they should not be used at the expense of the animal's need to breathe clean, fresh air from which the moisture content has not been dried out. If and when heat lamps are used, therefore, one must be sure to arrange adequate ventilation to make up for their drying effect. Lightweight, well-fitting clothing will otherwise best provide any extra warmth required, and leg bandages, applied not too tightly over cotton wool wadding, will keep the lower legs warm during enforced standing in the stable. In countries where heat can be a problem during the day the animal is often transferred to a shaded area on some hard standing onto which a hose-pipe can be played to cool the atmosphere.

When a horse or pony is feeling the cold its ears will feel very cold to the touch, and the hairs along its back and sides will 'stand on end'. This is referred to as a staring coat, and it is nature's way of helping to retain body heat. Air trapped between the raised hairs acts as an insulating barrier to prevent heat loss. (Humans still retain the mechanism of this skin reaction with goose-pimples.) It is well known that layers of air trapped between several lightweight garments gives much better insulation than one solid, heavy covering, so the lesson here is obvious. One must also be careful to ensure that an animal does not become overheated and start to sweat if a sudden rise in the external temperature takes place—the difference between daytime and night time temperatures, for example. It will be of great advantage to have two sets of clothing, so that replacements can be worn while the used rugs are shaken out and aired.

At grass

If a sick animal is not suffering from an ailment that makes stabling desirable, a period at grass when the weather is favourable will provide the ideal nursing environment. The horse or pony will benefit from plenty of fresh air and sunlight, which will greatly assist its return to health. The natural food and exercise will also assist healing and will have a beneficial effect on the working of the bowels.

For owners wanting to keep their horse or pony out at other times the system practised by a friend of mine is well worth considering. He has his horse turned out at grass while giving it free access to its stable direct from the field. Feeding and watering is done in the stable and afterwards the horse is free to go out and graze at will. In winter the animal is partly clipped, leaving the coat unclipped on its quarters and legs, and wears a New Zealand rug. On days when it is to be hacked or hunted it is brought in for the mud to be removed from its legs and neck and then it is lightly groomed all over before being ridden. In this way the animal is kept fit, healthy and happy and is worked regularly throughout the year; if necessary it can be left for periods without being ridden with no ill effects. Extra feed is provided in cold weather, when the grass is poor in winter or when the horse is working harder, and is cut back in spring or when it is not being ridden. This system keeps management work and time down to a minimum but ensures a high degree of care for the animal. A word of warning should be given about feeding time in the open stable. The animal soon learns to come for his feed and will often gallop in from the field. On one occasion this particular horse could not stop and went straight into the stable, where it skidded and slipped right over onto its side on the concrete floor. Luckily no damage was done, but now ther door is closed when the feed arrives before the horse can gallop in.

Feeding programme

It is well worth taking enough time and trouble to arrive at the proper amounts and varieties of feedstuffs that suit each individual horse or pony. I dealt with this in some detail in the first book of this series, (*Horsekeeping—Ownership, Stabling and Feeding*) but there are some points worth enlarging upon. The basic foods of oats, barley and bran fed with good hay in my opinion form a better feed programme for a stabled horse or pony than proprietary foodstuffs such as nuts and cubes or 'complete' feeds. When you use 'proper' ingredients you know exactly what is being fed and can control the horse's diet precisely. With manufactured feedstuffs you can never be quite sure what has gone into the feed or whether one batch differs in content from another. Manufacturers do, of course, take considerable trouble to ensure that their products contain the correct proportions of listed ingredients, but market variations in the raw materials mean that they may have to substitute one for another. There has been a growing number of incidents of 'mystery' ailments in horses and ponies fed on manufactured products. Although the number affected is proportionately small when measured against the very large numbers of horses and ponies fed in this way, it is nevertheless worth keeping in mind if you intend to increase the amounts of these products being fed to an animal or are changing over to them. Some owners have reported horses becoming unmanageable for no apparent reason, while others have had specific problems such as laminitis. It seems that horses and ponies not working hard are those most affected, particularly those being prepared for showing whose owners want them to look in fat-stock condition. Personally I find it better to err on the side of under-feeding concentrates rather than over-feeding them. It is very easy to increase up to the correct level without any trouble, whereas over-feeding will almost always cause some problems. In my view even carrying out the manufacturers' instructions for feeding their products can often lead to over-feeding. The amounts recommended can be too high, and although some manufacturers point out the need for owners to be mindful of this many fail to do so. It is tempting to think that by giving that little bit extra, or at least feeding the amounts suggested at the top end of the scale, one will ensure that one's horse or pony is better off. This is of course not so.

A more worrying problem is that difficulties can arise even when an owner has been feeding a manufactured product without any change in the amounts or regularity of diet. If he finds that the animal suddenly develops health probelms, there is good reason to suppose that they are the result of some variation in the ingredients. Manufacturers will never admit to this, but many owners have gone back to feeding 'true' foodstuffs like oats, barley and bran because they believe it to be true. Feeding concentrates in the manner I describe in *Horsekeeping—Ownership, Stabling and Feeding*, using crushed barley, oats and bran, a consistent feed programme can be devised and variations can be made with full knowledge of what is actually being fed. I use a combination of crushed barley and crushed oats, and find that by increasing the proportion of oats to barley only when work is increased any tendency the animal may have to 'heat up' on oats is cancelled by the extra work. Likewise, when the work is cut down or is of a less strenuous nature the proportion of oats to barley is decreased. It is also very easy to cut back just to crushed barley if the horse suddenly has to be rested, as its system is already used to the barley and in relatively small amounts this grain will do no harm to the animal when it is not being worked. Oats, on the other hand, should not be fed to a sick or resting animal; they have a higher level of toxicity that has adverse effects on the horse's system when it is not in work.

Appetite

When sick most of us show an initial tendency to be 'off' our food, and the same applies to horses and ponies. Don't let this worry you; no harm will be done if for a day or so food consumption is somewhat less than usual. Make sure to take away uneaten food, as this will quickly become unpalatable and will not be eaten if left in front of the animal. Do see to it that there is ample clean water, which must be kept fresh and untainted. To encourage the horse's appetite, prepare the food as temptingly as possible with mixtures containing known favourites. Generally speaking, all concentrates and grain feeds will have to be stopped or greatly reduced, and good hay and mashes relied on. Boiled barley, mixed with some crushed barley and dry bran, can be fed, together with sliced carrots. Bran mashes will make a good evening feed and will keep the bowels working properly; remember to add some salt to them, and slices of apple or carrot will make them very tempting to most animals even when they are not keen to eat.

Temperature

Your horse or pony's temperature can be a useful guide to its health when it appears to be off colour. In order to know exactly what is normal for your particular animal, it is a good idea to obtain from your vet a thermometer suitable for animals and use this to record the animal's temperature in health. The temperature of warm-blooded animals is remarkably constant in individuals, but there are normal minor changes during the day in most species. The temperature in a healthy horse should not vary by more than a maximum of 1.5°F (say 0.75°C) over twenty-four hours, 1°F (0.5°C) being about normal. It is usually higher in the evening than in the morning, and it will be higher after meals or exercise. (Digestion raises the temperature by approximately 1°F (0.5°C) and during sleep it drops by 1°F (0.5°C).) In pregnant mares and in young stock it can also be slightly higher. By taking the temperature we measure the heat of the blood, and with the horse this is usually done by the insertion of the thermometer in the rectum. Temperature can be taken at other places, but the rectum is by far the easiest place and will give a reliable reading of the internal temperature of the body. The normal temperature for a healthy adult horse is about 100°F, (37.8°C) varying from 99°F (37.2°C) to 101°F (38.3°C). It will be 1°F (0.5°C) more in very young or very old animals. Mares will be 1°F (0.5°C) less than male horses; the temperature of a mare in season will be about 2°F (1°C) higher than normal. Thoroughbreds will have a consistantly higher normal temperature than common-bred types, and violent work, especially in countries with very hot climates, will give rise to a higher temperature—sometimes by as much as 5°F (2.8°C) in a healthy animal. Prolonged exposure to cold and rain causes a lowering of the temperature.

The average temperature in health is 1°F (0.5°C) either side of 100°F (37.8°C); a rise of more than 2°F (1°C) is not compatible with health unless it can be attributed to one of the foregoing reasons. A rise in temperature of 2.5°F (1.5°C) indicates a mild fever; a rise of 2.5°F (1.5°C) to 4.5°F (2.5°C) a moderate fever; of 4.5°F (2.5°C) to 6°F (3.3°C) a high fever; more than 6°F (3.3°C) a very high fever. A temperature that is much lower than normal indicates anaemia or haemorrhage.

When the temperature is to be taken the thermometer should first be shaken to get the mercury down, and it can be smeared with a little vaseline or similar lubricant. To insert it, pull the animal's tail to one side and gently push it into the centre of the anus, where it should be held firmly for about two minutes. After use rinse it in cold water and dry it with a clean cloth.

2 Managing feedstuffs

If you buy your foodstuffs from a reputable merchant and in small enough quantities to be used up fairly quickly you should have no problems of food deterioration or loss of feed value because of age. You must still make sure, however, that you keep it in dry, vermin-proof conditions (as fully explained in the first book of this series). It is often much cheaper to buy crushed oats and barley direct from a local farmer; it may also be more convenient to do so if you have a supplier nearby who is willing to sell you small quantities of, say, 25 or 50 kilos at the time. Hay and straw purchases can also be made direct from local farms and once you have located a satisfactory source obtaining a regular supply will prove beneficial in several ways. Remember, though, that farmers are busy people, they will not appreciate being pestered for very small amounts or being expected to drop everything to attend to your needs. Give your supplier some advance warning of your requirements so that he can fit you in with his other work. Usually all that is necessary is a telephone call, a day in advance for grain, and with about a week's notice if you need hay or straw. Fetch small quantities of foodstuffs yourself, and if you want the farmer to deliver to you try to arrange it so that he can bring a full load. It is also wise not to keep him waiting for his money: if you pay him on delivery he will be more willing to come again. If you regularly use the same farmer and a load of hay happens to contain some bad bales; (perhaps very dusty or mildewed) do not ring him up and complain bitterly. If you explain politely that some of the hay is bad and invite him to come and see for himself he will often take your word for it and will just send you some replacement bales. Making good hay is not easy, and some bales of an otherwise good batch may heat up or turn mildewy.

They can also deteriorate in storage, and this is something to think about when you come to store it yourself. Hay and straw bales must be kept not only under cover but also raised up off the ground if the bottom layer of bales is not to be spoiled. Although many farmers do store hay and straw directly on the ground, knowing that the bottom bales will be lost, this is not practical when small quantities are involved. Bales are best kept up off the ground on some sort of staging, as even on a dry dirt or concrete floor the bottom layer of hay will quickly become unusable because of mildew. If you have no means of keeping it up, put a layer of straw bales on the bottom; as they are coarser and allow more air to penetrate, they spoil less quickly than hay. Straw is also cheaper than hay, so if you leave it long enough to become unusable you will lose less money.

An inexpensive method of keeping bales high and dry is to lay poles across small piles of concrete blocks. This will of course reduce the effective height of the storage building, which will have to be allowed for in your initial plans for storage facilities. Hay and straw keep best when the store has an open side, and this will also facilitate easy unloading. Keep the open side away from the prevailing wind, which will drive in rain, as wet and weathered hay will be wasted. For this reason provide the store with as big a roof overhang as possible.

Building a barn

Building a hay store or small barn need not prove difficult. I built my own quite large barn with only my father-in-law to help, and it was surprisingly easy. For those wanting to do likewise I will explain my method and materials.

First I engaged a contractor to level off a large area of waste ground. While he was doing this work he also took out the holes for the uprights, approximately 2 ft 6 in (75 cm) deep. A trench of similar depth was taken out, leading away to a convenient soakaway. Land drains were then laid and covered with old plastic sacks, and then filled in after first providing a drain opening for the downpipe from the intended barn roof. The bottom of each post-hole was filled with 6 inches (15 cm) of concrete, and when this had set 5-inch (12 cm) diameter poles were erected and the holes filled in. The poles were about 15 feet (4.5 m) long, and after going into the ground this left 13 feet (4 m) for the front height of the barn. The height at the open front end was actually more when the barn was finished because of the pitch of the roof and the overhang at the front. For the rear uprights I used second-hand railway timbers 10 feet (3 m) long, giving a height above ground of 8 feet (2.5 m). These uprights were arranged to provide four bays each 10 feet (3 m) wide.

I next notched out the top of the front poles. 7-inch (17 cm) x 2-inch (5 cm) wooden beams were lifted first onto the top of the rear railway timbers and then at the front to fit onto the notched-out poles. They were bolted through the poles and fixed down to the rear timber supports, and angle irons made from old bed frames

Building a barn
It is important to construct the barn floor well clear of the ground: store the bottom bales to allow air to circulate: failure to do this will result in bales becoming spoilt by mould and mildew.

The barn roof should overhang for weather protection on the open side but be securely fastened down against the wind.

Roof overhang
secured against wind

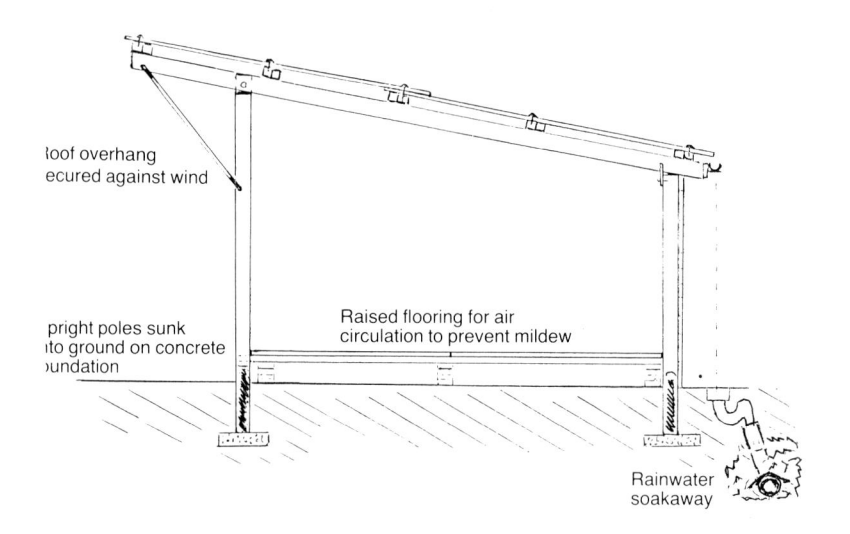

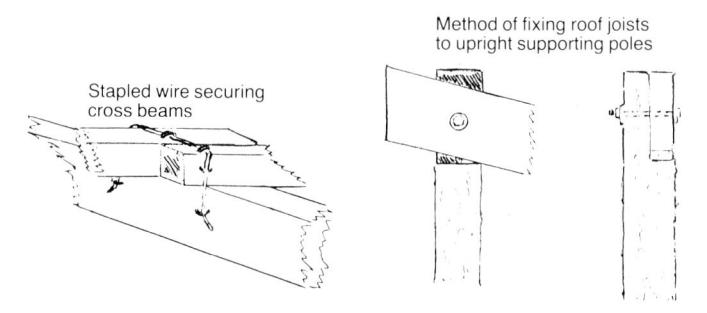

Upright poles sunk
into ground on concrete
foundation

Raised flooring for air
circulation to prevent mildew

Rainwater
soakaway

Stapled wire securing
cross beams

Method of fixing roof joists
to upright supporting poles

Fig. 1 Details in the construction of a barn.

were nailed from the front overhang of the beams to the poles for extra support. Finally I spanned each bay with 3in x 3in (7cm x 7cm) timbers and covered the top with galvanized metal sheeting. Before this was fitted I stapled heavy gauge wire around each roof timber to the beams in order to prevent strong gusting winds from lifting the roof off. The sides of the barn were then clad in a mixture of second-hand timber and galvanized sheeting, and the whole thing was completed by the two of us in less than a week.

 Before using the barn to store hay and straw I built staging some 2 ft 6 in (75 cm) off the dirt floor and then, to be sure of keeping my hay in good condition, I covered the staging with a layer of straw bales, putting the hay bales on top of these. When I have mixed deliveries of straw and hay I arrange for the straw to come on the top of the load so that it can be off-loaded first and then the hay put on top. This makes for practical management as I can use the

Method of notching out the top of the supporting poles to carry the beams of the barn roof.

combination of hay and straw by taking it out in lines across the barn as I need it.

Grain storage

Nothing beats the galvanized metal containers that are specially made for grain storage, but they are expensive. If you cannot afford these bins you will need to provide other means for storing your grain and other concentrates to keep it free from interference by vermin. Without such protection, even if your feed store seems vermin-proof you will undoubtedly have vermin trouble, with bags being eaten through and your grain being spoilt or wasted. A stable cat (if it is not overfed on pet food) will be a great help, but you will have to shut it in the store to sleep at night: it is no good having the cat stretched out in front of your fire on cold nights while the vermin are eating your costly horse or pony feed.

Without a cat or the galvanized bins I favour, other methods of protective storage will have to be sought, and I have heard of people using old chest-type deep freezers for this purpose. The idea sounds ingenious, but I have not tried it myself. My only slight reservation is that they are completely airtight, and I wonder whether grains might not tend to heat up in them, though prising off the rubber seal might overcome this. The lids of the metal bins always allow a slight air gap around the flat lid, but are still vermin-proof because of their smooth sides; it should not be too difficult to make this provision with an old deep freeze.

Vitamin and mineral supplements

These can play an important part in the feeding programme, especially of a stabled horse or pony without access to grass. Many horsekeepers maintain that if the correct amounts of good fresh feedstuffs are being fed it is not necessary to give supplements containing extra vitamins. This may be true, but the difficulty even when feeding natural foods is to be sure that some of the vitamin value has not been lost by drying and storing. For this reason I believe a vitamin supplement to be desirable. Because I take considerable trouble to ensure as far as I can that my feedstuffs have retained most of their vitamin value, I only add a supplement of half the amount that is recommended by the manufacturer. In this way I do not waste money on a lot of unnecessary extra vitamins, but I ensure that should there at any time be a small deficiency some supplement is present to make it up. I find that this system works admirably. If you do use supplements, do not be tempted to mix them as this can cause toxicity and imbalances.

The same applies to minerals. Minerals are taken up from plants, which have taken them from the soil. There are some fifteen essential minerals, all of which will be present in a vitamin–mineral supplement of the type offered by leading manufacturers. The average owner need not, however, worry too much about the correct balance of the mineral intake as with balanced feeding this will take care of itself. One important factor that should be remembered, however, is the balance of calcium to phosphorous. Twice as much calcium to phosphorous is the approximate requirement estimated for horses or ponies, and it is unwise to exceed this proportion. Too much phosphorous is harmful and can lead to obscure bone problems and lameness. Because of its relatively high phosphorous content, it is therefore unwise to feed a horse or pony on large quantities of bran alone; it would give far too much phosphorous and too little calcium, with the resulting harmful effects. Do not, however, be frightened to use bran in the diet as prescribed, as it is a very useful foodstuff and most beneficial when used in correct proportions to other feedstuffs.

Horses and ponies will sometimes be seen to eat dirt or paw the ground to expose roots and then eat these. This is a sign that the animal is trying to obtain minerals that it knows by instinct it requires. A stabled animal will sometimes also eat straw in an effort to obtain alkaline to relieve acidity in the stomach or bowel, possibly caused by a build-up of worms. A horse or pony not usually given to eating its straw should have its feeding and management programme examined if it suddenly begins to do so. Occasionally, straw bales contain succulent pieces of dried grass, and this will cause the animal to ferret about in the straw; if you are sure this is the reason, the habit can be ignored as of no consequence.

Salt licks

Another essential part of the diet of any horse or pony is salt. This is necessary to enable the body to utilize protein and so avoid exhaustion and fatigue. The best way to make sure that your horse or pony gets enought salt is to place a salt lick in the bottom of the manger, so the animals can lick or nibble at it whenever it wants. A combined salt lick, which contains minerals such as iron, cobalt, manganese, zinc and iodine, etc., and is usually about the size of a small brick is ideal for use in this way. My own horses can often be heard both day and night going to their mangers and pushing their salt bricks about to lick them or turn them over to find any small pieces of feed stuck on the underside. This also helps to keep them amused for short periods without any harmful vices being learned from it. I also add a small amount of salt to each evening feed, especially in hot weather or when the animal has had reason to sweat because of hard work.

3 Digestion and digestive disturbances

Care of the teeth

Most digestive disturbances will be avoided if one's management and feeding is carried out along proper lines. There are, however, one or two things that can occur however much care is taken to feed the correct amounts and types of food and not to overload the system of a tired or sick animal. It may be that your horse or pony quite suddenly and for no apparent reason refuses to eat its food, or perhaps when it does feed you notice that it has difficulty in chewing and in doing so holds its head to one side. Another sign that something is wrong is if it dribbles or the food drops out of its mouth in small balls (called quidding) when chewing. These can all be signs that there is trouble in the animal's mouth, and you should immediately check its teeth. If left unattended the poor creature will not only be miserable but will lose condition as a result.

The examination should begin with the molar teeth. These are the two rows of large teeth along each side of the jaw, top and bottom, located behind the bars of the mouth where the bit rests. Because of the side-to-side chewing action of the jaw the tables or grinding surfaces of the molar teeth gradually wear uneven. Eventually a razor-sharp edge can result along the outside edges of the molars in the top of the jaw and the inside edges of those in the lower jaw (called 'hooks'). The molars naturally close together with sloping tables and this leads to sharpened hooks as wear takes place. Sharpness on the inside of the lower molars can irritate and cause cuts or ulcers on the tongue; sharpness to the outer edges of the upper molars will probably cause the inside of the cheeks to be similarly affected.

When this happens the sharp edges will have to be rasped off, and it is usual for the vet, or sometimes your blacksmith, to do this. There is no real reason why you should not carry out this operation yourself provided you can obtain the proper type of rasp of the correct size to fit your horse or pony's teeth. These rasps are specially designed to fit over the row of molar teeth, and have a

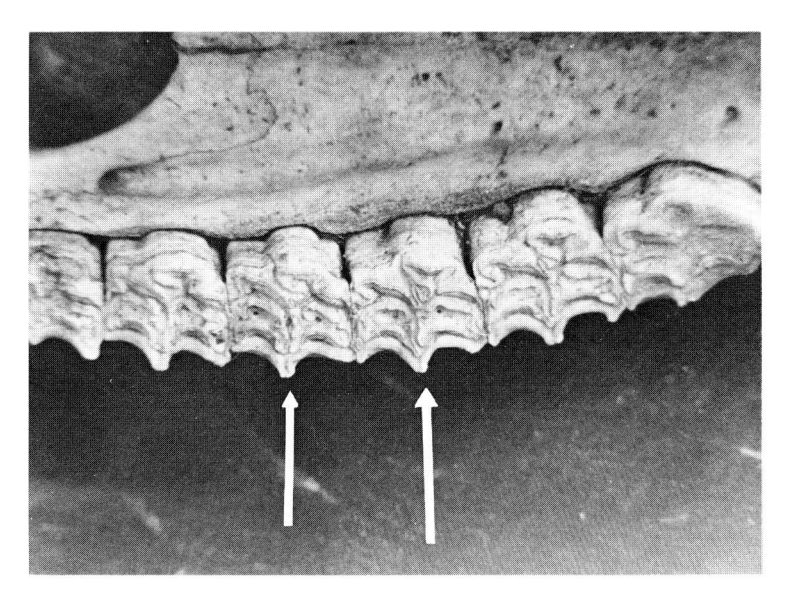

Skeleton of jaw (viewed from the inside), showing upper molar teeth with very pronounced 'hooks' to their outside edges.

smooth outside surface so as not to injure the inside of the animal's mouth when used. They are very expensive, however, so not worth the outlay unless you have a lot of horses. It is best to have two people when checking a horse or pony's teeth for sharpness, and when any rasping is carried out two people will certainly be necessary even when the use of a gag is employed. (Gags are metal instruments that when placed in the animal's mouth and adjusted make it impossible for the mouth to be closed.) If the animal is haltered it makes it possible for one person to carry out the rasping while the assistant holds the animal's head. However, I dislike the use of gags, as in my experience these often make an otherwise placid horse or pony object strongly to the procedure and thereby

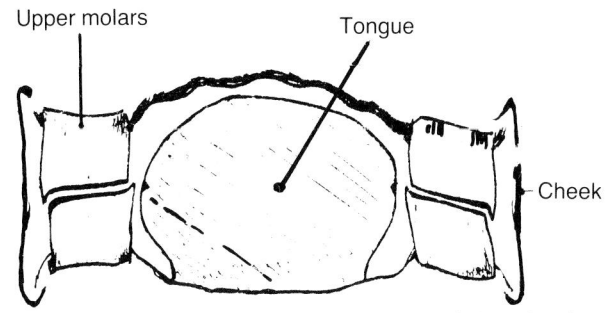

Fig. 2 The angle of molar teeth and the development of sharp hooks.

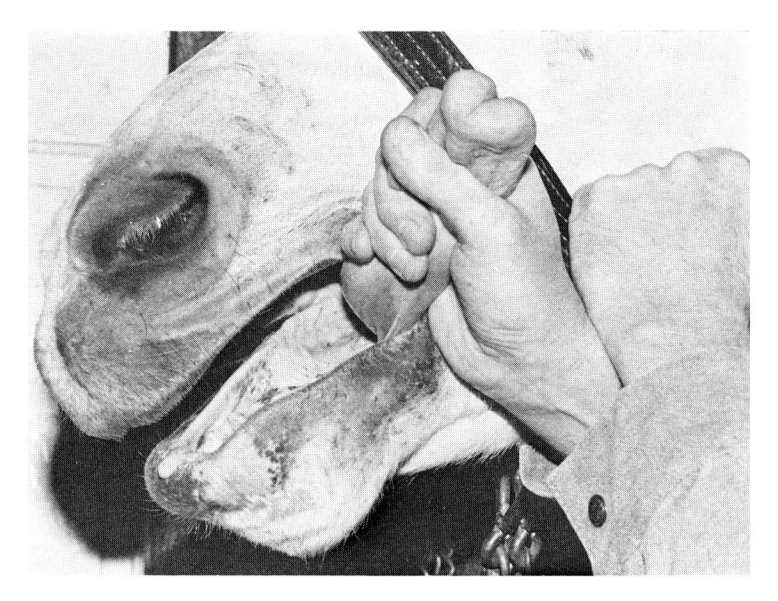

Method of holding the tongue to one side in order to inspect or rasp the molar teeth on the opposite side.

make the job more difficult. Certainly a nervous horse is likely to become extremely fractious. An alternative method for an examination or rasping of the molar teeth is to withdraw the animal's tongue to the side of the mouth and then have this held there by an assistant. It is best to place the horse or pony with its hindquarters in a corner of the box before you begin. Warm the rasp in a bucket of warm water before rasping as the animal will then usually object to it less. Use this water to rinse off the particles of teeth that collect in the rasp after several strokes. Rasping should be continued *only* until the fingers can feel that the sharp edges have been removed. Do not rasp away the tables or their natural angle. The outside edges of the upper molars usually require more attention than the inside edges of the lower ones. If the animal becomes very fidgety then the assistant may have to resort to the use of a twitch. Intervals between the necessity for rasping will vary from horse to horse; it is a good idea to examine the teeth for excessive sharpness every six months or so. The front incisor teeth are never rasped.

Correct salivation

The importance of chewing properly, and so mixing large quantities of saliva with the food before it is passed on to the stomach, cannot be over-emphasized. Horses and ponies should always be

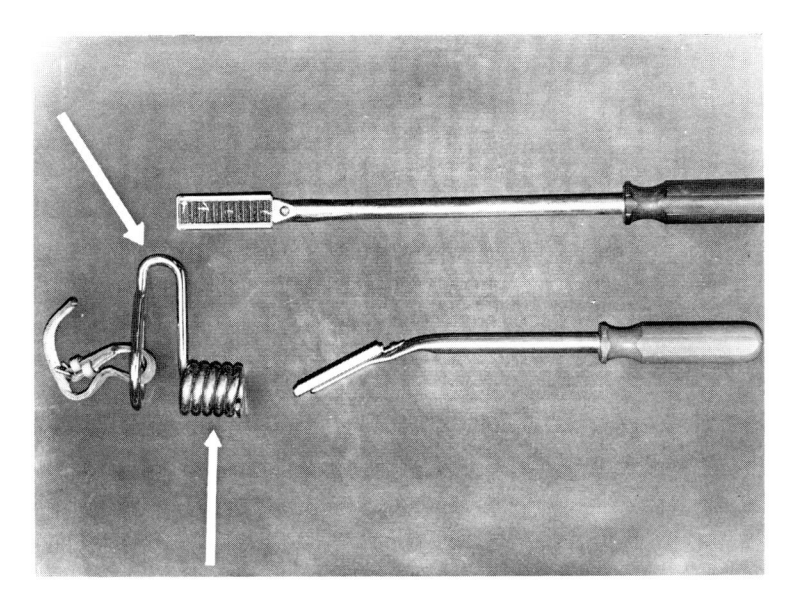

Special rasps for removing the sharp edges of the molar teeth: note that the rasping area does not extend to the outside limits of the instrument to avoid damage to the inside of the cheeks and gums. The angled (lower) rasp is used for the upper molars. The gag fits into the mouth with the coiled part (bottom arrow) between the molar teeth; the animal's cheek fits into the looped part (top arrow) and the strap can be used to fix it to a headcollar and hold it in place. This is called a Swales gag. With difficult horses the strap is often left undone so the gag can be removed quickly if the animal becomes fractious.

encouraged to eat slowly by being fed without too big a gap between meals. Do not over-damp feed as this will encourage the horse to bolt the meal. A good way to slow down a greedy eater is to place a salt brick in the manger: this will prevent large mouthfuls being gobbled down too quickly. Remember, too, that when mixed with saliva and digestive juices large feeds of concentrates will greatly increase the volume of bulk in the stomach. In excess this food will stretch the stomach and some will remain there for too long, where it will ferment. Gases will be produced and cause pain (colic), and the resulting increase in acidity will be absorbed into the blood. This can lead to loss of sodium, which becomes excreted by the kidneys with the excess acids. Eventually an imbalance of body fluids will result and the animal becomes lethargic and later develops muscular tremors culminating in azoturia or laminitis. It must also be appreciated that the breaking down and digestion of food is greatly assisted by necessary bacteria, which are present in the digestive tract in very large numbers, especially in the caecum and colon. They live and reproduce in this environment, and any sudden changes or greatly unbalanced feeding will adversely affect their well-being.

The feeding of hay provides the basis of good dietary management for the stabled horse or pony. By its very nature hay has to be eaten slowly, with the addition of much saliva that is produced with its chewing. Properly made hay will not ferment, and it passes from the stomach quickly to be digested farther along the digestive tract. The addition of balanced concentrates, fed three times a day in accordance with work being performed, will then give the horse or pony all the extra energy and maintenance requirements it needs. If concentrates are fed in smallish amounts after the animal has first eaten some hay as a 'starter' to mobilize its saliva and digestive juices, no troubles of the kind described will be experienced.

Diarrhoea

This condition is usually temporary and originates from one of a number of causes: sudden changes in diet, especially onto green fodder; excess of foods of rather high fat content, such as barley; heavy worm infestation; feeding of grain that has become 'heated'; scouring of foals due to the mare's foaling heat; exhaustion through overworking an unfit horse or pony. Good management plays a part in prevention of the condition (except in the case of foals), but should it occur through any of these causes then the remedy is obvious. Any sudden change in feeding will upset the sensitive stomach of the horse or pony, especially of those that are finely bred. Even when care is taken with an animal brought in from grass and gradually introduced to stable feeding with concentrates, diarrhoea will sometimes occur for a few days before the horse adjusts to the new routine. This may be considered as normal and is of no consequence.

Constipation

This is most commonly caused by a poor diet or an excess of concentrated foodstuffs. It may also occur after the use of purgatives if the system was severely purged. Again the remedy is to look to management. Bran mashes and/or a period at grass will usually be the best treatment, and particular attention should be paid to correctly balanced feeding.

Colic

Colic is the word loosely given to describe evidence of pain in the general region of a horse or pony's abdomen. It is more common in horses or ponies than in any other animal, mainly because of the complicated structure of their digestive tract. The comparative smallness of the stomach can lead to food passing into the intestines without being adequately mixed with digestive juices, and the twisted up arrangement of the intestines positively invites food to lodge there. A horse or pony cannot vomit as most other animals can, and this means that the cause of the trouble has to stay put until it can be cleared right through the system. There are many causes of colic, many of them the result of bad management in the form of incorrect feeding. Numbered among them are the following: too long an interval between feeds; sudden changes in diet; too much feed for too little work; feed of poor quality, such as 'heated' grains or mouldy hay; feeding too soon after hard work; allowing excessive drinking of cold water when returning hot; inadequate supply of water over a long period; teeth troubles leading to insufficient mastication; feeding of grain or hay that is too new or green. Other causes may be accidental access to grain or concentrates; eating straw bedding; foreign bodies in the feedstuff; poisonous plants in hay or meadows. Another common cause when management has been neglected is a build-up of worms. Periods of the year when flies are bad can often be the cause of colic, as many horses and ponies will gallop about or become so frenzied that they upset their digestive functions, causing gas to be produced and retained in the tract.

Symptoms that colic is present are a general uneasiness or restlessness; frequent turning of the head to look at the stomach; attempts to kick the sides and stomach; attempting to pass water or droppings without doing so; the horse attempting to get down and then changing its mind, or getting down and up again quickly. When any of these signs are present and colic is suspected, lead the animal quietly around the yard or paddock and then put it into a loose-box of adequate size for free movement. It should be well bedded to prevent injury should the animal get down. It is considered by some to be very important not to allow the horse or pony to roll because of the danger of a twisted gut, whereas others fear no such occurrence. Generally speaking I think it best if the animal can be kept standing. All food should be withdrawn, but access to clean water allowed. This is the only treatment I advise to be given by the layman. If it is not serious the condition will pass within a couple of hours. If it does not, or if the symptoms become more violent, then summon a vet immediately. Vets will treat such a call as an emergency; they will well know of the possible serious causes that may need their prompt and expert attention. Many experienced horsekeepers will give the animal a drench with something like a 'Brown Draught' and then wait for an hour or two for signs of improvement before summoning a vet. This can be dangerous, however, as many of these drenches are a poison and so the original cause of colic can be made worse.

Worms

Infestation from parasites will lead to a horse or pony losing condition, and if left untreated will result in many ailments and disorders, including anaemia and intestinal problems. Horses and ponies that develop large bellies, look unthrifty, are listless and whose dung has a highly offensive smell should be suspected of having a large infestation of worms.

There are several types of worms that infest horses and ponies. The most common, and the one that gives rise to the most trouble, is the red worm or strongyle (*Strongylus equinus*). A fit and healthy horse or pony can play host to a certain number of worms; in fact these will invariably be present to some degree. A harmful build-up will occur when horses and ponies are over-stocked and graze on horse-sick pastures. The reason for this is twofold: first, because under such conditions many parasites will be present and will multiply rapidly, and second, these conditions produce an acid reaction in the digestive tract which forms an ideal environment for the worm cycle. Serious infestation of red worms leads to inflammation of the bowel, which can actually become punctured. If the worms then work their way to the main arteries supplying the gut and penetrate these they will cause a blockage; once the blood supply is cut off or severely restricted in this way acute colic will result and a painful death to the animal will be the consequence.

Management in the form of pasture control by 'resting' and other measures, as described in my previous book, will constitute the main preventative action. Regular dosing of the horse or pony with worming powders or paste will also be necessary. I find that powders are the easiest to use. It is advisable to use a selection of such treatments so that the parasites do not build up a resistance to any one type. Frequency of dosing will vary with conditions but will probably be from once every four to five weeks when a build-up is present to once every six months for normal dosing. With horses or ponies who will not eat their food if it is 'tainted' with worming powders the modern disposable syringe that squirts the worming paste down the back of the throat is useful.

Lung worms

These are rare; when they occur they affect the bronchial passages of the lungs. Donkeys are the most natural hosts for these parasites, and horses or ponies that are turned out with a donkey so infected can themselves become infected through the faeces. If a horse that has been in such contact develops a cough, the disease should be suspected and faecal samples should be taken by your vet. If you intend to share a paddock where a donkey is being kept it is also a good idea to have this examination carried out before you agree to turn your horse or pony out with it.

Azoturia

Azoturia is something of a mystery disease. It used to be known as 'Monday morning sickness' because many working horses were affected by it when beginning work on Mondays after being rested over the week-end. Shortly after exercise begins there will be signs of muscle stiffness and pain mainly affecting the hindquarters. This can be so severe that the animal will lie down and be unable to move, and in very extreme cases kidney damage and death have resulted. A closely related disease, often called azoturia or tying-up syndrome, affects very fit horses being fed on high levels of concentrated feed. They will suddenly show symptoms of pain and stiffness with a reluctance to move, and the muscles of the hindquarters will become hard. Sweating and tenseness may also be present. It can often occur in these animals when their exertions from exercise have finished and they are cooling off. The treatment in both cases is to get the animal into a loose-box on good bedding as quickly as possible. With true azoturia the animal will often go down and be unable to rise, and will often fail to urinate. The affected muscles should be massaged and veterinary help summoned immediately, as specialized treatment in the form of injections will be needed. It is also often advised for the horse to be given two taplespoons of bircarbonate of soda four or five times a day to help neutralise the acidity in the blood; this is administered by opening the horse's mouth and depositing the powder on the back of the tongue. If the trouble is merely that of being 'tied up', then with massage and movement the animal will often recover quite quickly and the attack will pass off without complications.

Both these conditions result from digestive disturbances and an imbalance that can be caused by incorrect feeding, either by overfeeding in relation to work being done (or not done), or by an excess of various nutrients which the animal's system cannot cope with properly. Prevention will be the result of proper dietary management and an understanding of the individual's feeding requirements. Many problems—laminitis and colic, bone disorders and unsoundness as well as azoturia—will be avoided if the ratio of diet to exercise is understood and properly carried out. Many animals have been ruined or had their working lives cut short because of the ignorance of owners who have consistently fed them incorrectly. With modern horse and pony management these problems are still with us, and in fact are on the increase. Azoturia and similar disorders are affecting many quite ordinary animals, not just those in hard work, and I believe this is in many cases the result of the horse or pony having too much of a good thing. Manufactured foodstuffs and supplements, especially when mixed together with grains and other compounds, can disrupt the vital nutritional balance. At first this may result in an unwillingness to work or the horse or pony being difficult to handle. If this is not recognized as being due to dietary imbalances it will eventually lead to specific disorders. Imbalanced or over-feeding leads to the storage of excess glucose in the muscles, and particularly in large muscle masses such as the hindquarters. When the horse is subjected to exercise or work, especially if it is not 'warmed' into it gradually, the muscles become deficient in oxygen and the excessive glucose converts to lactic acid, producing a large amount of toxin. This disrupts the normal metabolic working of the muscles, which are then unable to function properly. Pain and muscle damage result, giving rise to the symptoms already described. It is important, therefore, not to overfeed or to give large single feeds of concentrates that the animal's system cannot cope with at one go; nor should various compounded feeds be mixed together. When changing over to a different foodstuff make sure to introduce it gradually over a period of two or three weeks. Never bring your horse or pony in from fast work or when fatigued or very hot and feed it concentrates or let it cool off suddenly. Always cool it down and let it unwind properly if you do not want it to 'tie-up'. Another thing to beware of is turning it out onto grass that has been recently fertilized as this has also been known to give rise to these problems.

Some interesting work has recently been done on the treatment and prevention of azoturia and similar disorders by the use of vitamin E. It has been found that vitamin E on its own is of no use if it is accompanied by a selenium deficiency, as without this mineral the vitamin E cannot be taken up by the body; low selenium leads to low vitamin E utilization. Tests have shown that certain parts of the world (and certain parts of different countries) have a selenium deficiency in the soil; this has been further aggravated by the overuse of chemical fertilizers, etc. A balanced intake of selenium plus vitamin E, fed as a supplement, is claimed to produce very good results in the prevention of azoturia in these areas. It has also been found beneficial in the treatment of joint troubles and arthritis in older horses.

4 Wounds and skin disorders

With any injury suffered by a living creature, be it a horse or pony or a human, the actual treatment administered never cures the patient. What it does is to assist the body to affect its own cure by removing as many of the obstacles as possible from the location of the damage so that natural healing can take place. We use treatment to this end so that natural healing is speeded up and the intervening period of discomfort to the patient is reduced. There are certain fundamental requirements needed for all patients recovering from illness or injury, and most people will be aware of these. Warmth, good clean air, cleanliness and the right sort of tempting food are those common to all convalescents, as are rest and light exercise. There are only a very few exceptions to this general rule, such as the need to apply coldness to a localized area to reduce swelling or the initial withholding of food in the case of colic. The more we can induce a feeling of comfort and normalize the body's temperature, the more quickly the patient will respond and return to a healthy state. If horse and pony owners pay heed to these basic needs then much suffering will be avoided and many additional vets' bills will be saved. Remember this—neglectful management to save you time today will cost you time and money tomorrow.

Bruises

When the body suffers a blow the resulting contusion when the skin is left intact is what we commonly know as a bruise. The damage suffered is to the underlying tissues of the skin, and to some extent the nerve cells also suffer injury. A haemorrhage occurs within the skin and soft parts and this gives rise to swelling and heat due to the increase in blood together with a coloration (or discoloration) of the area. This in turn produces pain and immobility. These simple bruises will disappear of their own accord, and rarely need treatment other than rest and time for nature to complete the healing process. Sometimes when a blow of great intensity is sustained, or if it is struck on parts of the body not shielded by thick

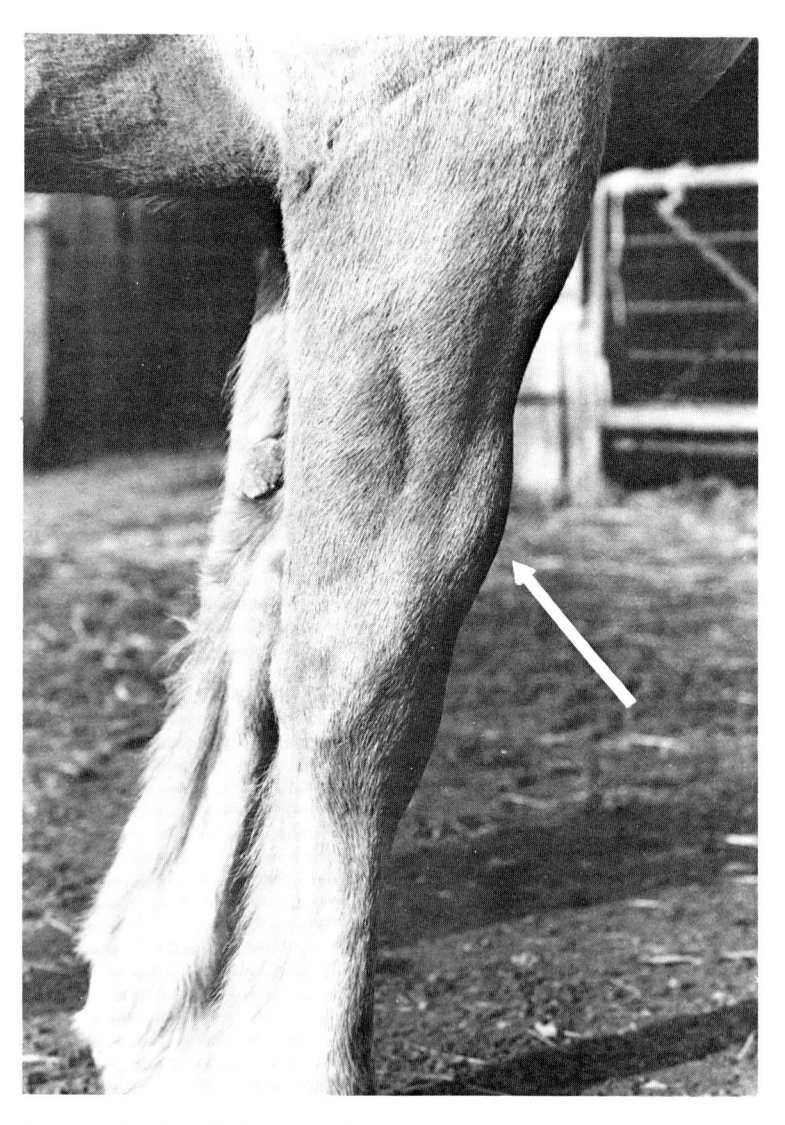

An example of the kind of lump that, though a blemish, is of no detriment to the animal's fitness; in this case a permanent enlargement probably caused by a blow to the extensor muscle.

layers of muscle, the penetration of damage will be deeper. When it is to the lower leg region of a horse or pony, for instance, the underlying tendons, synovial sheaths and sometimes the actual bone will be affected. In this case treatment will follow the lines described later in this book. But for simple bruising the first task will be to decrease blood flow to the area, and then to reverse this process and increase the rate of blood flow. This is because the immediate reaction to a severe blow when the skin remains unbroken is for inflammation to occur where the blood vessels are ruptured. The natural reaction from the body is threefold: to effect a repair by bringing cell-building materials to the site; to remove the clotted blood and damaged tissues by carrying these off in the bloodstream; to control and arrest the internal bleeding. These functions are carried out simultaneously. Swelling and heat occur at the site because of this increase in blood and fluid and also because of that escaping as a result of the damaged tissue. What we must therefore attempt to do to assist natural healing is to reduce the size of the damaged blood vessels that are leaking blood in order to make repair easier, and at the same time to increase the size of the blood vessels that are bringing the repair materials so that healing may be speeded up. This is impossible, so we must try to compromise. If, for example, the area of the tendons is in a bruised state and the injury is noticed immediately so that early treatment can be given, we first rest the horse to decrease the flow rate of the blood and then use a cold application to the affected area to constrict the blood vessels and reduce bleeding from the damaged parts. This treatment is carried out frequently during the first twenty-four hours. Each cold application should be kept to a few minutes so that the temperature of the limb can quickly neutralize the coldness and no prolonged chilling takes place. Devitalization of the surrounding tissue can occur if we go so far as to sustain a very cold temperature to the area (taken to the extreme, frostbite can result). In between this cooling treatment we need to control the fluid from filling the area and so the leg is bandaged. The area is covered with gamgee or similar dressing, which is bound into place, working from the bottom upwards to drive the fluid up the leg to be carried away by the circulation above the knee. The pressure of the bandage will also restrict the escape of more fluid and maintain compression of the damaged blood vessels.

Carried out over the first twenty-four hours this short-duration cooling together with the bandaging will reduce and control subcutaneous bleeding, keeping the bruise to a minimum and so shortening the time needed for its repair. From the next day after this we can set about the task of helping nature to do the actual healing, and that means warmth to the area and massage to stimulate blood flow. Bandaging can still be applied between treatments, this time to keep the area warm.

Wounds
When an open wound is suffered there are additional complications because infection will be present due to the breaking of the defence

mechanism provided by the skin. The way will be open for invading germs to attack the patient's body, and in this case the safest course is to summon veterinary aid, especially if the horse or pony has not recently been given protection against tetanus. (If the wound involves the tendon sheath or a joint then veterinary help must be summoned at once.) If you decide, however, that the wound is not of sufficient proportions to warrant calling the vet, you will need to administer treatment different from that described for bruising.

Bleeding from the wound will in the first instance be doing good, as it will wash away infection to some extent and help to prevent germs from gaining easy access. Running cold water over the area for a minute or two will also help. Then clean the wound with a saline solution (one level tablespoon of salt to a pint (550ml) of water), removing any pieces of debris stuck to the wound, and dab it dry with a clean piece of gauze. In fairly minor wounds natural clotting will now have begun and bleeding will become less pronounced. (Very marked bleeding would need to be controlled in the first instance by the use of pressure pads.) Once the wound has been cleaned it can be covered with an antiseptic ointment and either bandaged or left open at the discretion of the attendant and depending on the location. It is obviously better to cover a wound that is likely to come into contact with dirt or other foreign agents. The body's natural reaction to open wounds will, as with bruising, be to repair the damage, but it will also include the task of fighting any invading germs. When infection takes place these germs will have multiplied in a few hours and will be penetrating deeper. Treatment must therefore be aimed at assisting the body to send increasing numbers of white blood cells to the affected area, where they will attack, kill off and ingest the offending organisms. These dead cells will then be pushed to the surface of the wound by copious amounts of serum that has been released to clean them out. (This is pus.) One result of this activity is intense inflammation, and we must help to relieve this and at the same time help to stimulate blood flow to the area. Hot applications will be helpful, as they will dilate the blood vessels, speeding up the migration of white blood cells, and also stimulate the lymph exudation of the area. (This is also why hot poultices are used where swellings need to be brought to a head.)

Specific advice about the treatment of wounds is impossible without seeing individual cases, but the foregoing explanations and guidelines should enable an owner to recognize what treatment is needed or when veterinary assistance is required. Generally speaking I think it advisable for a vet to be summoned whenever damage to tendons or joints is suspected, and also for punctured wounds because of their tendancy to close over, trapping a deep-seated infection. With long, open wounds stitching will be necessary, and it should be remembered that with wounds to the lower limbs of horses and ponies there is the problem of the formation of excessive tissue, causing proud flesh.

Skin disorders

Skin troubles of various kinds often get neglected as they seem to be of minor importance, but in fact once the barrier of skin protection is broken the way is left open for bacteria to invade the body and set up infection. What starts as a seemingly unimportant lump or cut can, if neglected, soon become a painful wound that will often take a long time to heal. These areas can be particularly troublesome if they occur where tack is worn or where the skin is constantly stretched when movement takes place.

Sore backs

These can be very tiresome if an otherwise fit horse or pony is prevented from working. Useful management practices will avoid your horse or pony being off work because of a sore back. Always ensure that the saddle and girth fit properly and are correctly adjusted. Clean and check them regularly to ensure that there are no stiff or rough edges that will rub the animal. When saddling up, especially in winter, never slap a cold saddle straight onto the horse or pony's back, and everything you place on its back (including yourself) should be lowered on as gradually as possible so that the animal's muscles have a chance to adjust to the weight and maintain its balance. Imagine the horse's back as a plank suspended between two chairs: you will appreciate the need for not suddenliy jolting a heavy weight onto the middle of it. In winter a warm numnah placed on the back before introducing the saddle is a desirable practice, and in summer a thin saddle cloth is also advantageous. These seemingly little things are of more importance than many people realize, and failure to think about them is often the unwitting cause for many a horse or pony refusing to be tacked up or bucking once the rider gets on.

In summer, when a numnah under the saddle would cause excessive sweating, I have found a cotton saddle cloth a very useful item. My wife makes these from old cotton sheets by simply cutting a rectangle and stitching around the four sides. A cloth fitted beneath the saddle prevents any friction from the saddle on to the animal's back and also absorbs much of the sweat that would otherwise be absorbed by the saddle's leather panel. It also keeps the saddle clean. Make sure that the cloth is large enough to leave an inch or so (some 3 cm) showing below the saddle panel. The cloths are easily washed and dried after use, and when worn or torn can finish their life as stable rubbers or rags.

When returning from exercise never remove the saddle

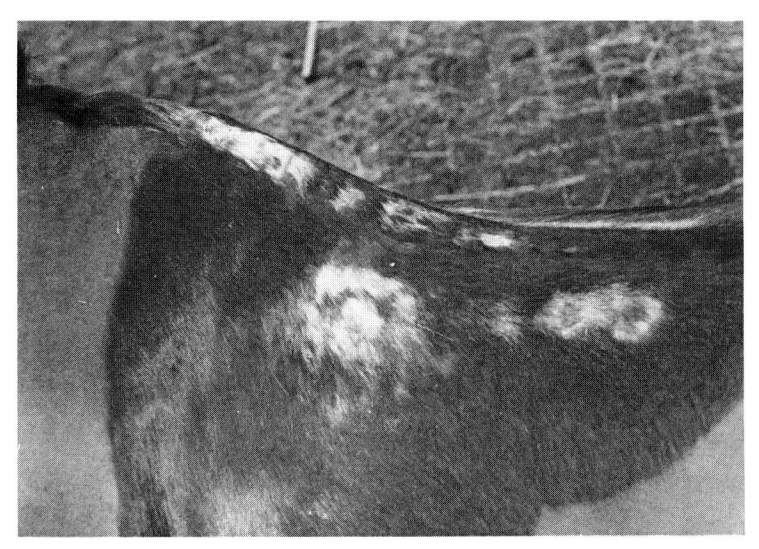

Sore back
The type of sore back that can result if the advice given in the text is ignored.

immediately but loosen it and leave it in place for several minutes while you remove the bridle and halter the horse or pony. When you do remove the saddle cover the patch with an old towel to allow cooling of the skin to take place slowly. This will prevent the animal's back from being 'scalded' by sudden exposure to the air while the pores of the skin are open. Blisters can result if this precaution is not taken, and if the horse or pony is ridden in this condition it will result in ruptures of the small blood vessels of the skin, and a painful lump or even raw patches will occur. Badly fitting or over-tight girths can also produce this condition, known in this instance as girth galls. In order to restore full circulation to the skin where the pressure of the saddle has restricted it, always massage the back after it has cooled off by firm slapping with flat hands. After exercise or long periods of work the horse or pony's neck and hindquarters will also benefit from this form of massage, as it will help to relieve tired muscles by stimulating fresh blood supplies. Do not be too heavy-handed though, especially to the area of the loins. Another practice to avoid is to ride an animal for too long a period if it has recently been brought in from grass after a long period without work. In this soft condition its back and girth area will be particularly susceptable to blistering from saddle and girth pressure. If you keep these things in mind and carry out the procedures described, your horse or pony will never suffer from the discomfort of a sore back and you will not have an otherwise fit animal that you cannot ride.

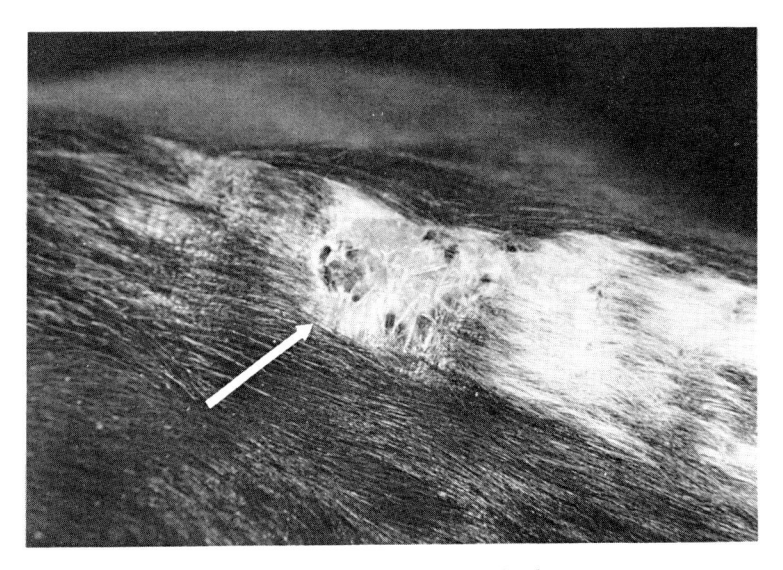

Close up showing the scab formation as the skin heals.

In my view a horse or pony with a sore back should never be ridden. The use of a numnah or something similar with holes cut out to fit over any sore spots to prevent the saddle from pressing directly upon them is sometimes recommended but I consider this to be bad practice; the pads often move or become squashed down in use and are then useless—in fact they can add considerably to the animal's discomfort. Should there be any lump or bare spot to which the horse or pony shows pain by flinching to even the slightest pressure of the fingers then do not try to ride until the condition clears up. This may seem very disappointing at the time, but to do so will invariably lead to a much longer forced lay-off. I have, however, seen animals with hard, bare areas or callous lumps from old wounds in the saddle area that have been ridden without apparent discomfort. On the owner's assurance that her horse had these permanent places on its back as a result of old scars I saw one such animal passed fit for a long-distance ride. When it returned I was interested to see if the condition had worsened as a result of the competition, and to my surprise and relief I found on examination that the condition was unchanged and the animal showed no sign of distress when the area was subjected to pressure. It goes to show that there are times when discretion can be used, but the risk should never be taken with lumps or wounds that are raw or show any sign of discharge. The treatment for such cases is not to ride the animal and to clean and lightly dress the affected area with a soothing antiseptic cream until healing takes place.

Another reason for lumps to appear under the saddle is the constant use of an unclean numnah, which will cause an area of skin to be rubbed or infected so that a painful lump is produced. This can also happen when numnahs or rugs and blankets have been washed with a liquid cleaning agent that has then not been properly rinsed out and causes a skin reaction. Such seemingly minor lumps can become very troublesome, and in spite of care and treatment often remain only partially healed. They may eventually result in a form of skin tumour, which will be a constant source of pain and irritation to the animal when it is ridden. The vet can often remove these small lumps surgically, and the skin will then heal over to leave a clean, smooth area that will allow riding to take place after a month with no recurrence of the problem.

Warbles

Another problem not so easily avoided, especially with horses and ponies kept at grass is that produced by the warble fly. These flies lay their eggs on horses' legs, or in long grass where they are then collected onto the animals' legs when grazing; when a horse or pony eats the grass or licks its legs to relieve an itch the eggs are ingested into the system. This takes place in the late summer. Nothing is seen until the following spring, when the larvae that have hatched out within the horse's body make their way out to the skin. The horse's back and sides are most often affected, the larvae being 'drawn' by the saddle. A small, hard lump will appear underneath the skin. Within ten days or so the swelling will enlarge and produce a 'head', where a small hole will appear as the larva is ejected. It is most important not to apply pressure to the lump in any form (i.e. riding or squeezing) until the larva is ready to come out: doing so will make it burst, causing infection over a much larger area, which will persist for weeks throughout which time the animal will be incapacitated. Treatment should be to try and hasten the 'ripening' of the larva, and thus its eviction from the skin, by hot fermentation or poultices to soften the skin. When the hole in the lump is seen to be large enough for the larva to be ejected, slight pressure with the fingers may be used to bring it out. Do not squeeze too hard to try and force it out if it is not ready, as it must not burst inside the skin. Also, do not use a poultice once the hole appears and starts to open, but continue to use hot formentations instead until it is ready. When the larva comes out there may also be some fluid and a little blood; after this the hole closes and soon heals and the swelling disappears. In some cases, and especially if the larva has been allowed to burst inside the skin, some permanent scar tissue will be left. If the affected area is not beneath the saddle or girth normal exercise may be continued.

Cracked heels and mud fever

These are two similar conditions that affect horses and ponies. Cracked heels, as the name implies, occur in the hollow of the pastern, whereas mud fever affects the limbs from the hock or knee downwards and also, in some cases, the abdomen and stifle. The condition is similar to that suffered by humans when wet hands get chapped by wind and the skin dries out and cracks open. In horses, infection in the form of dermatitis results. At first the skin will become reddened and tender; later, little scabs will form, the surrounding skin will die and then drop off, together with the hair. Movement (such as that at the back of the pastern) that stretches and bends the skin will cause it to crack and break open, thus allowing germs to enter, and more inflammation takes place. If left unattended the condition can last for weeks, and is very painful. When, as is common, the cracked heels occur at the back of the pastern on a hind leg the animal will often snatch its leg up as the tender part of the skin is stretched apart. The affected area can become quite swollen and the animal will walk on its toes, though the lameness disappears with exercise. The condition occurs mostly in wet, cold weather and is more prevalent in districts with acid soils or soils containing silicates, where the sharp crystals will penetrate the softened skin and allow infection to take place. An itchy skin will often give warning of the condition. Animals with white legs are more susceptible to this trouble, especially if the underlying skin is pink. Many people will refuse to buy a horse or pony with white feet if it is required to stand out in winter, because such animals are more likely to suffer than those with darker skins. because of the irritants found in mud it is generally advised not to wash it off when returning from exercise as the pores of the skin will be open and the irritants can be rubbed in and so cause trouble. Dry mud can safely be brushed off without it penetrating the pores, which will have closed during the drying period. With a stabled horse, however, I gently dry off the area behind the pasterns on returning, using an old towel, unless the legs are absolutely caked with mud. Bran is also useful when applied as a drying agent. Zinc ointment or vaseline applied to the skin behind the pasterns before beginning work is very helpful in preventing trouble, especially with pink-skinnned horses and ponies.

If the trouble has occurred, treatment is to keep the area clean and dry and use an application of zinc ointment or one of the antiseptic skin creams that are available—I find Protocon ointment very good. The legs should also be kept warm, and bandages worn in the stable will increase the blood circulation. Make sure that there are no cold draughts from the bottom of the stable doors. Oats are best withdrawn from the diet because of their heating effect and the resulting demands on the blood to clear toxic waste from the system. Bran mashes will be helpful in increasing toxic elimination from the bowel, which will relieve the skin of some of the work of waste disposal. Because dirt can penetrate the wound, make sure that the horse or pony has been given its anti-tetanus injections; with severe cases of cracked heels a course of antibiotics is very helpful.

Sweet itch

This is another infection of the skin, usually affecting the mane and tail area but sometimes extending down the neck to the withers. It is found in all types of horses and ponies, though its incidence is greater among ponies. It is associated with lush grass, and is often first noticed when the animal is seen rubbing itself to relieve itching when out at grass in the spring. The marked frequency of its occurrence in fat cobs and ponies while on pasture gives rise to the belief that certain plants have an allergic effect on the animal. This is probably true, and I also believe that itching is caused by the richness of some grasses eaten by ponies that would do better on less lush grazing being 'thrown out in the blood'. This is, however, only a partial explanation, as the condition is essentially a dermatitis attributed to dirt and neglectful management. When only irritation occurs, with no skin lesions, the possibility of lice should be investigated. Part the hair at the base of the mane and tail, which are favourite haunts for lice and nits, and closely examine these areas for any signs of their presence. If any are found, treat with sulphur ointment or lice powder and remember to disinfect all the grooming kit, blankets, etc.

When sweet itch is the cause of irritation the skin becomes scaly and may show small ulcers with a little pus. The hairs stick together and may break off near the roots, leaving moist, bare patches that later become hard and wrinkled. As sunlight is said to aggravate the condition and lush grazing is a contributory factor, the animal is best stabled. If this is not possible for each 24-hour period, try to keep it in at least for most of the day and turn it out at night, when it will eat less and there will be no sun. The affected areas should be clipped, washed with soap and water and then thoroughly dried. Calomine lotion, zinc ointment or one of the many skin creams can then be applied. If the condition does not respond to this treatment and proves difficult to cure the vet should be consulted with a view to a course of injections that will help.

Sweet itch can also start up in wet weather as a result of the softening effect of rain on the skin, but another form of dermatitis known as 'rain scald', could in these circumstances, be wrongly diagnosed as sweet itch.

Rain scald

Rain scald can affect the same areas as sweet itch, but it may be found anywhere on the upper parts of the body. It does not occur in dry weather and is most likely to be found during a spell of prolonged heavy rain. Although the condition is similar in appearance to sweet itch it is caused in this case by a species of fungus invading the skin. Most people consider no special treatment to be necessary; the condition is aggravated by exposure to continuous rainfall—the remedy for this is obvious.

Ringworm

This is a highly contagious disease affecting the skin caused by a fungus or, to be precise, many types of fungus. The horse is particularly susceptible to two varieties. The infection may be spread from horse to horse (or pony) or from contagion from stables, gates, tack, grooming kit or by humans acting as a carrier. The disease usually shows first on the forehead, face and neck, or perhaps at the root of the tail, but it can spread to all parts of the body. At first all that may be seen are what look like infected insect bites, and itching may or may not be present. After a few days, circular areas of hair become raised and matted, exuding a small amount of fluid. The hair falls out, leaving a bald patch covered with dry scales that are greyish in colour. Veterinary advice is needed about suitable treatment, which may be in the form of iodine ointment or an aniline or, more likely, an anti-mycotic such as Griseofulvin given as a feed additive. It may be difficult to be sure when the condition is cured, but when no further scaling takes place and the hair begins to grow it can usually be assumed that a cure has been effected. Once ringworm is suspected, isolate the animal and disinfect all walls, manger and stable equipment. Burn bedding and any other disposable equipment. Where there is no danger from fire the fungi can be destroyed from walls, etc., with a naked flame such as a blowlamp. Keep the cleaning kit and tack of each animal separate, and clean them after use with a powerful solution of antiseptic. It is also advisable to wear rubber gloves when grooming an infected animal.

Warts and sarcoids

Warts are quite common in horses and ponies and may appear on any part of the body. They generally occur around the head and neck region, another likely site is between the thighs. They take the form of outgrowths from the skin and vary in size: some of them become very large, sometimes reaching the size of a fist or even larger. They have been likened to a small cauliflower in appearance.

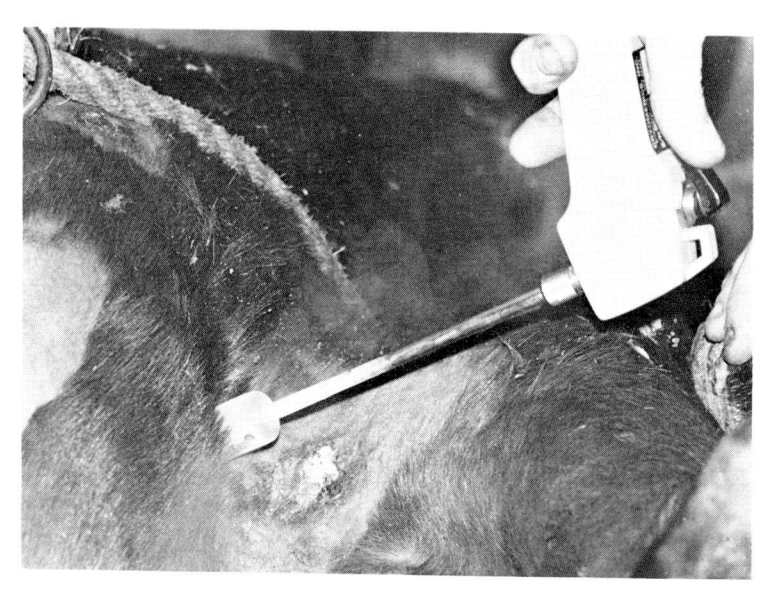

Sarcoids in the area of the groin being treated by cryosurgery. This produces liquified nitrous oxide through a narrow tube, causing intense freezing.

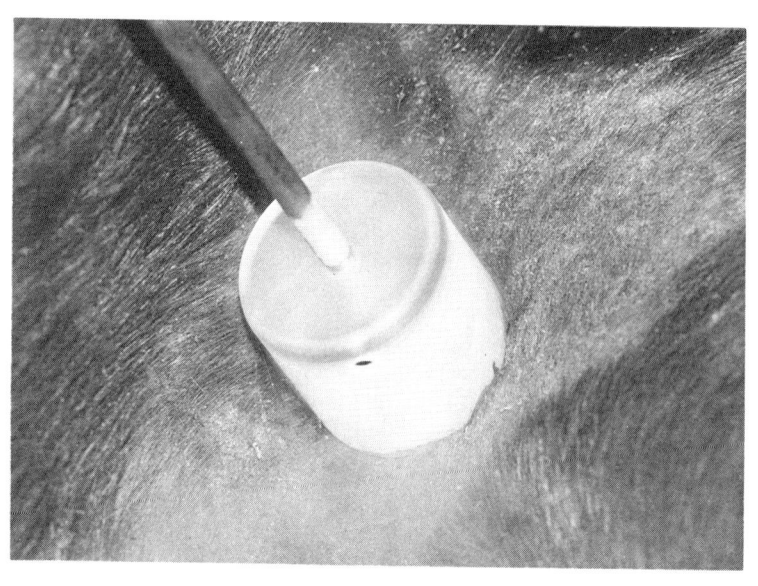

Close up showing the instrument completely enveloping a plaque-type sarcoid during freezing with nitrus oxide; this causes the malignant cells to burst and emit toxic fluid, thus destroying themselves.

Your veterinary surgeon will remove them and effect a permanent cure. Small ones (papilloma) that appear around the nose and head of young animals will often disappear if left alone.

Sarcoids are a much serious problem, as these growths are malignant and have a great tendency to recur after surgery. They appear as whitish lumps with grey or pink shades and two types exist: the round-celled type is the most malignant. The modern method of treatment for these growths is cryosurgery, which is the controlled use of freezing to destroy the unwanted tissue. Nitrous oxide instrumentation obtains a temperature of $-20°C$ ($-4°F$), and this is controlled to act on the site, causing tissue necrosis. The postoperative period produces odour, drainage and the eventual separation of the dead tissue, and this can be assisted by hydrogen peroxide soaks.

Melanoma

Finally, while on the subject of tumours, there is the melanoma, which is a densely pigmented lump found only in grey horses (very rarely in other colours). They do not occur until the animal is six or seven years old; it is estimated that some 80 per cent of greys over fifteen years are affected. The most frequent site for these tumours is the under-surface of the tail or the skin around the anus. Sometimes they will be seen on the head somewhere below the ears.

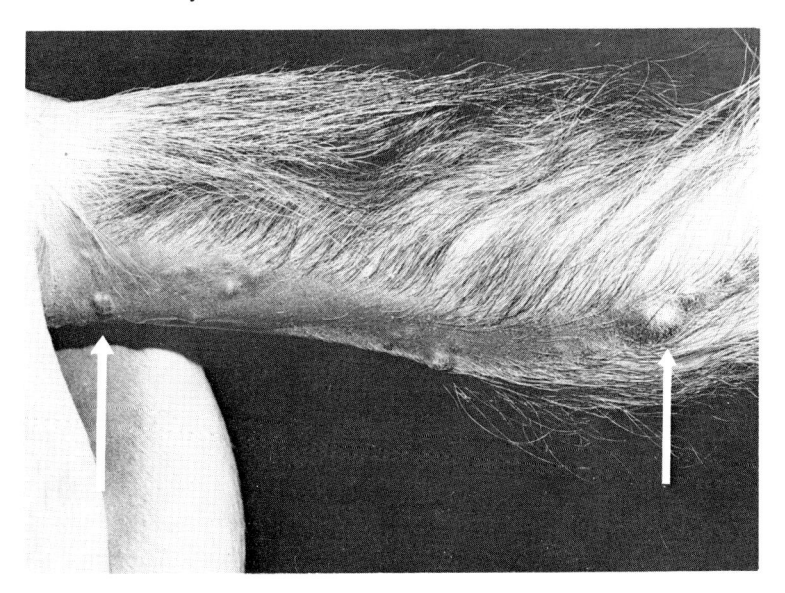

Melanomas on the underside of the dock. The primary can be seen (arrowed right) with many secondaries between it and the arrow left.

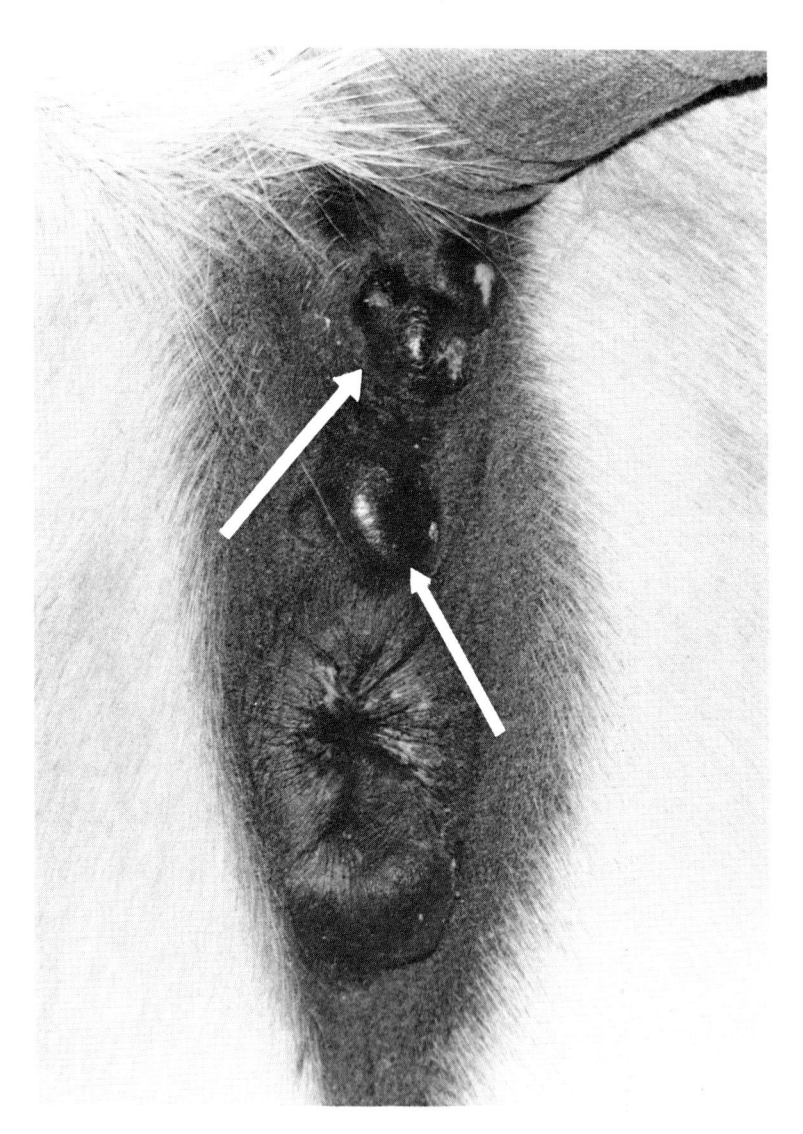

Typical melanomas found around the anus of many grey horses and ponies; these should be left alone.

They are well-defined, round in shape and quite firm. The reason that only grey horses are prone to them is said to be due to the natural loss of hair colour, which causes a process in which pigment is appropropriated to form a tumorous lump. They should be left alone, and care must be taken not to interfere with them during grooming. They are slow-growing and rarely give any trouble. I used to ride a grey mare that had a collection of these around the anus; she was still hunting well into her twenties and was well over thirty when she died.

Loss of pigment

To give it its proper name, 'leucoderma' is not a disease but a condition of the skin left after an injury or disease. Ringworm can be responsible for leaving white, pigment-free areas devoid of hair. Badly fitting saddles that cause rubbing can also leave such areas. Another place often affected is the sides of the mouth where a rubber bit protector has been fitted, causing the loss of pigmentation. There is no remedy for the condition once it has taken place.

Tetanus

This disease is caused by a germ gaining access to the body through a wound. A punctured wound where the skin closes over trapping the infection inside is particularly likely to be conductive of tetanus infection. Tetanus bacilli live in the soil and particularly that which is highly manured. A wound can become infected immediately or later by soil contamination; symptoms of the disease may only appear as much as 3 weeks after a wound had been inflicted. The affected animal will become stiff in its gait with difficulty in backing or turning and later muscle spasm occurs with sweating and increased respiration. The muscles of the jaw are affected so that the animal is unable to open its mouth (hence the name lockjaw); another symptom is the partial covering of the eye by the corner membrane.

Prompt veterinary treatment in the early stages of a mild attack can result in recovery but in acute cases death will result. The need for immunization with a regular yearly 'booster' dose is, therefore, very important for the horse or pony's welfare.

5 Coughs and colds

Simple coughs can be brought about by minor disturbances to the delicate membranes of a horse or pony's respiratory system by the inhalation of irritants which, if left to continue unheeded, will lead on to much serious trouble. A horse or pony's need for a plentiful supply of good clean air cannot be too greatly emphasized. The horse is particularly susceptible to respiratory ailments and problems affecting the nose, throat and lungs, and these conditions are often brought about by a continued lack of fresh air. A horse or pony out at grass, even in inclement weather, will seldom suffer respiratory troubles provided that it is not neglected by undernourishment or left to freeze in cold, wet weather. If, however, it is brought in from grass to a badly ventilated stable or one giving a sudden rise in temperature a cough or cold will quite often develop. The bedding may also be to blame, especially if it is old straw containing mould or spores, or is very dusty. My own preference for bedding is deep-littered wheat straw, properly managed by adequate daytime 'airing' and frequent skipping out of droppings while in use. Nothing but good, bright wheat straw totally free of spores or dust must be used to start off such a bed. It should be spread to a depth of sufficient thickness to provide a resilient covering over the whole floor area, and then be allowed to build up to a consolidated depth of 6 inches (15 cm) by daily topping up. This provides a warm, comfortable bed that can be left undisturbed for six months or even longer if the correct management programme is followed. (This proces was explained fully in the first book of this series.)

Colds and chills can also be caused by bringing an animal in from exercise too hot and then standing it in a stuffy or draughty stable to cool off. Always, therefore, walk the last mile or two home, depending on how hard the animal has worked and its stage of fitness. Once home, loosen the saddle but leave it in place while you check the feet for stones and injury. Remove the bridle, and while this is being cleaned leave your horse or pony haltered with the

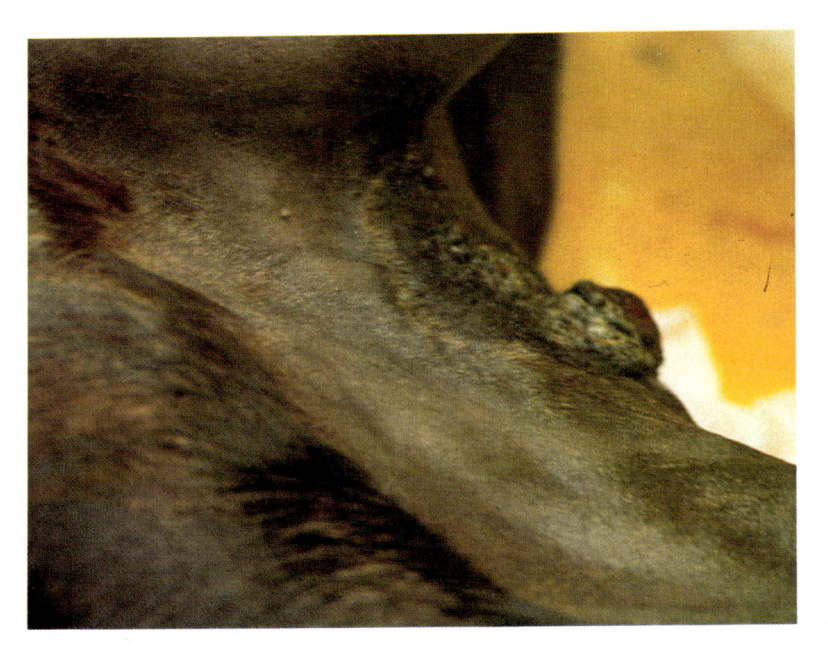

A sarcoid under the throat, showing evidence of secondaries spreading in the area.

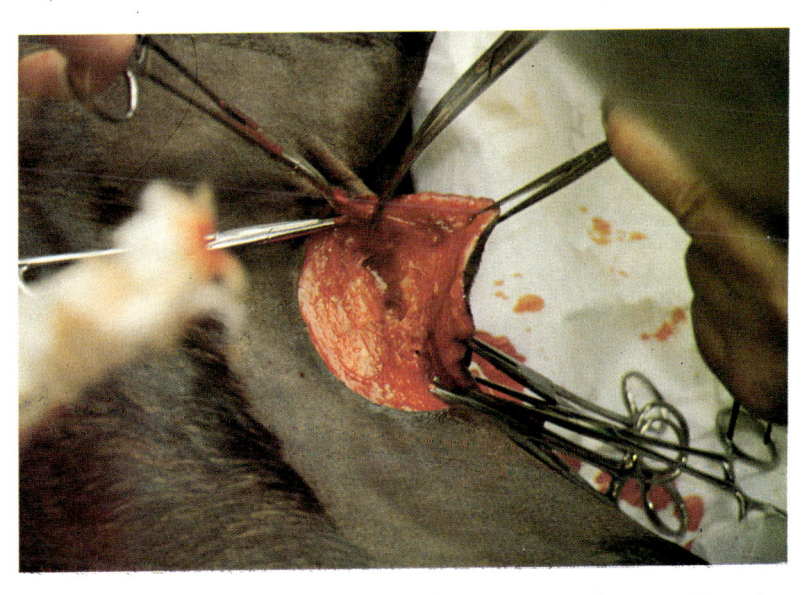

Surgical removal of the sarcoid was necessary as cryosurgery could not be attempted because of the closeness of the jugular.

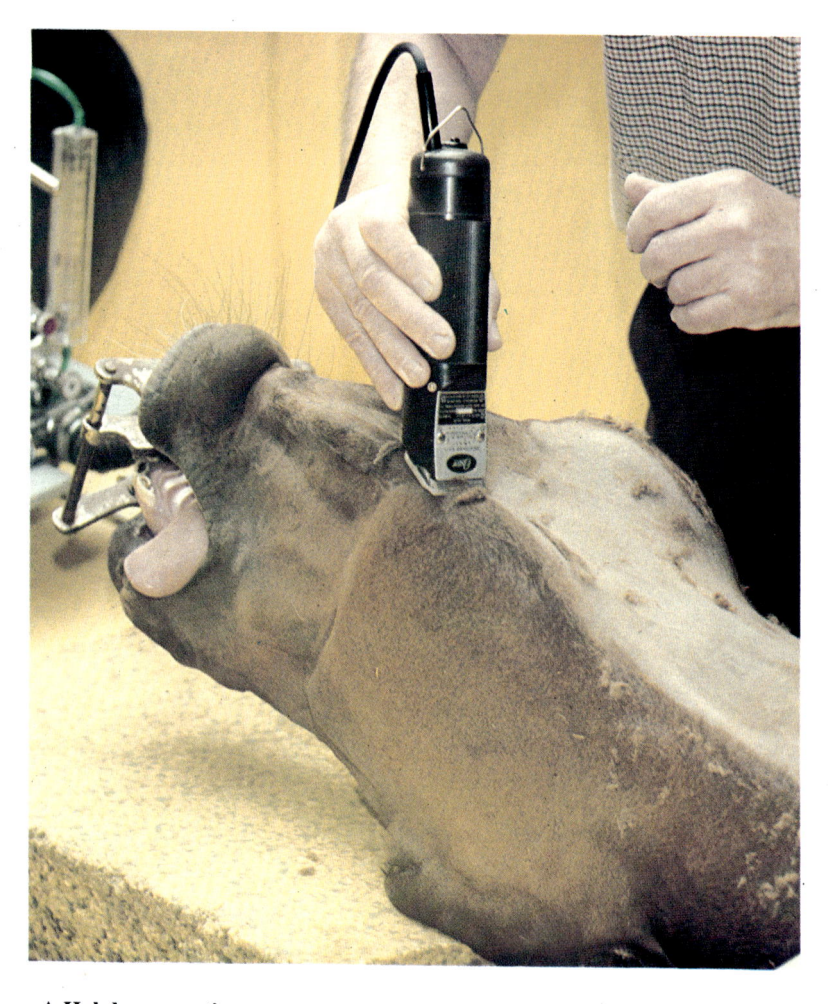

A Hobday operation
Preparation for the operation: the horse is fully anaesthetized and on closed circuit breathing apparatus.

The vocal cords of the larynx are exposed through the centre of the surgical opening. ➤

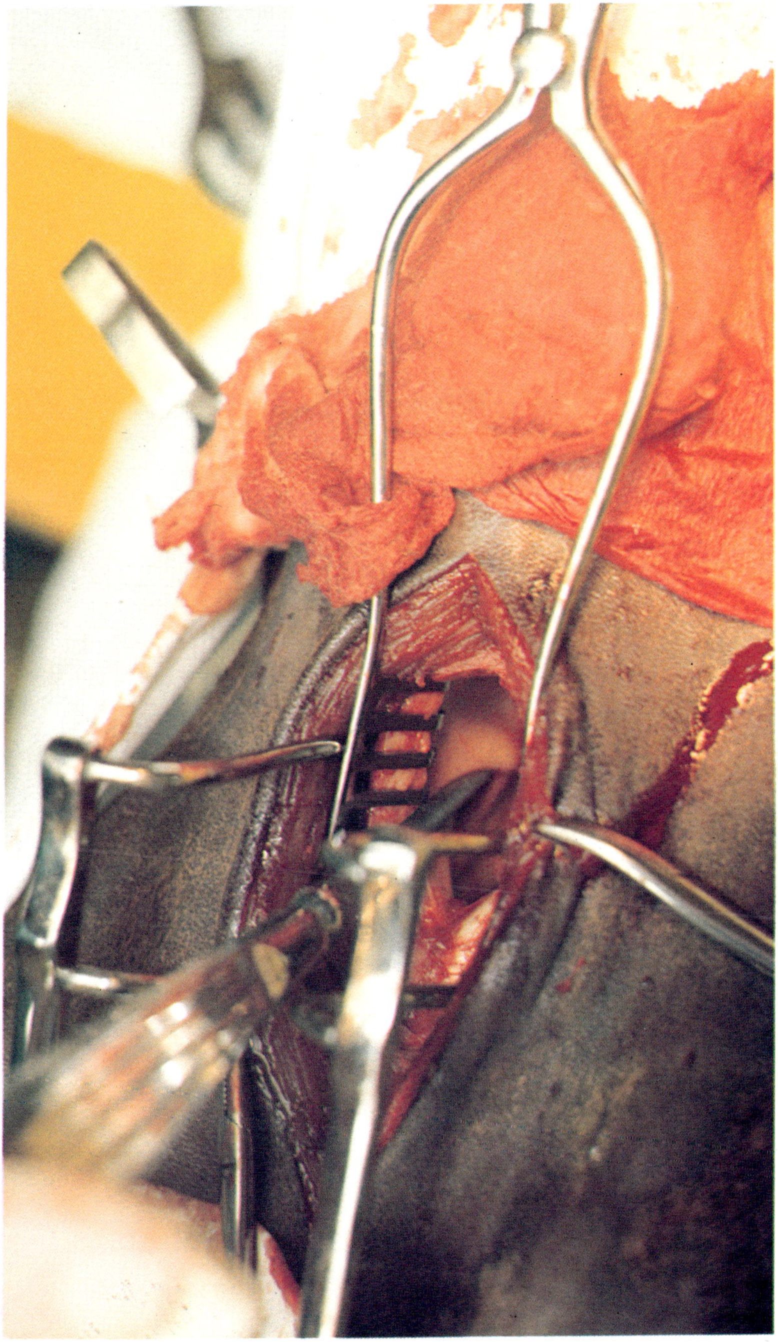

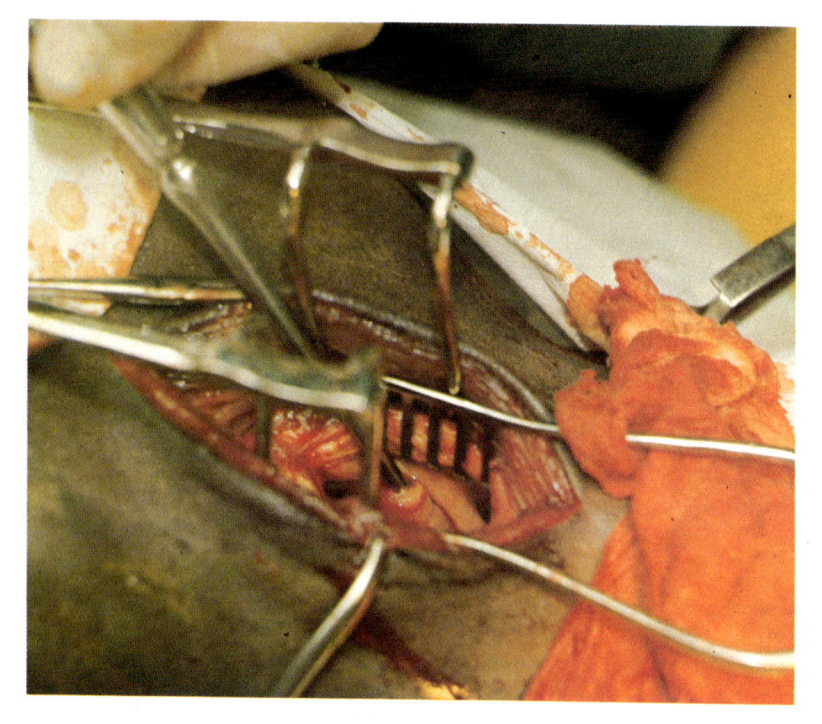

Surgical removal of the mucous membrane liner of the ventricle, known as ventricle stripping.

The removal of the membrane liner (centre picture) will cause a tightening of the vocal chord and thus prevent whistling and roaring. The operation is called a Hobday after the man who first perfected it.

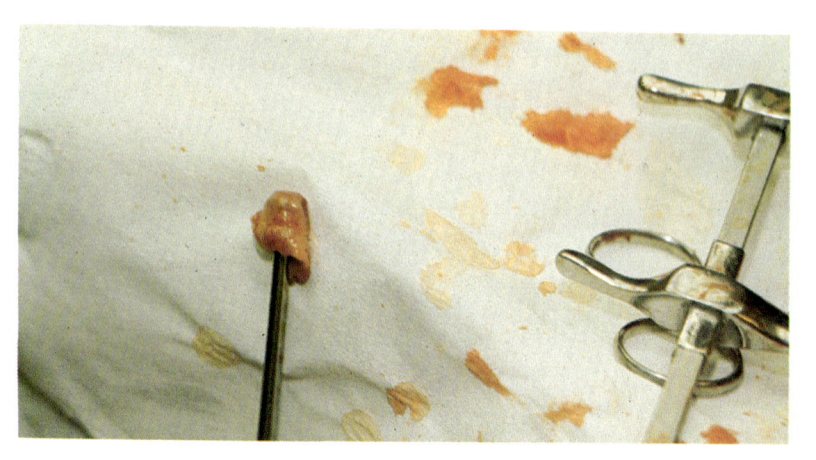

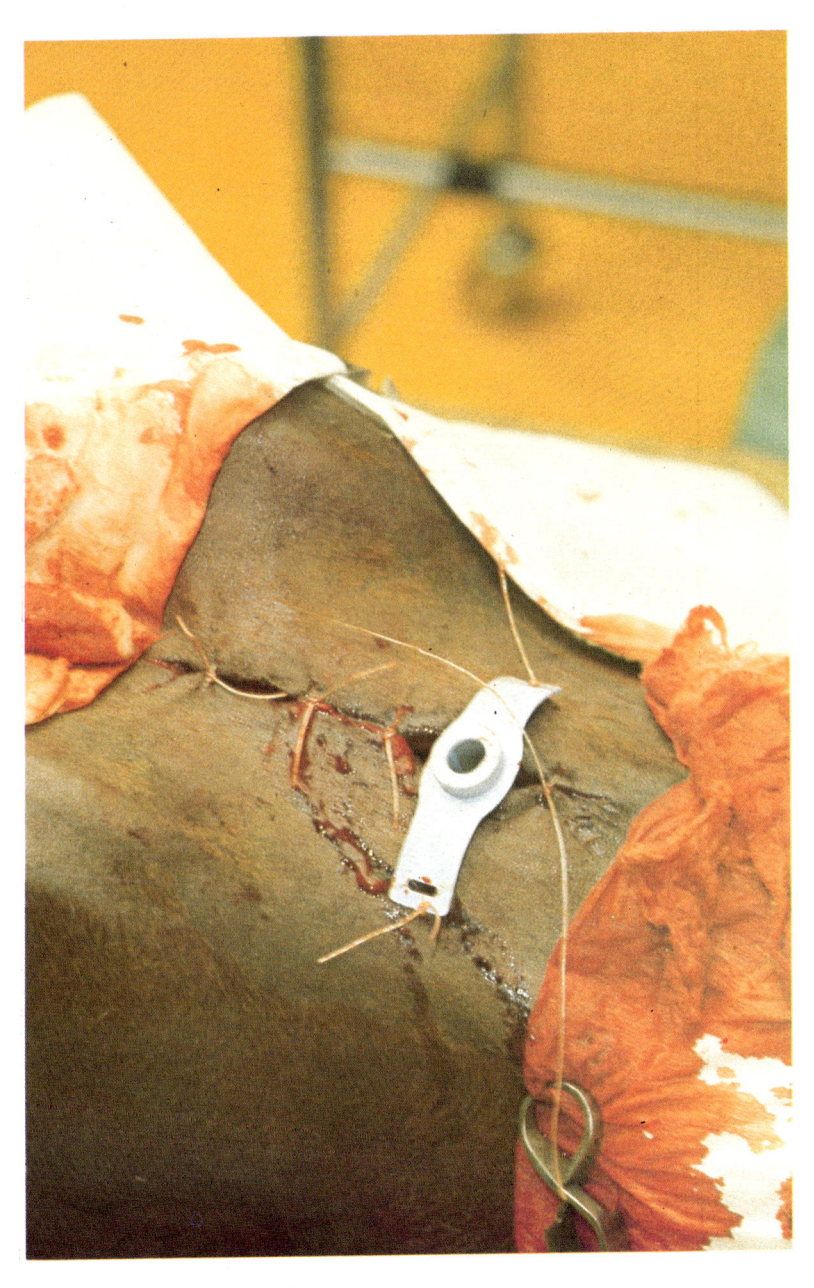

When the throat is stitched up a temporary plastic tube is inserted to allow free drainage on healing.

Line firing
Line firing being carried out by cauterization of the skin in the area of the back tendons of the foreleg.

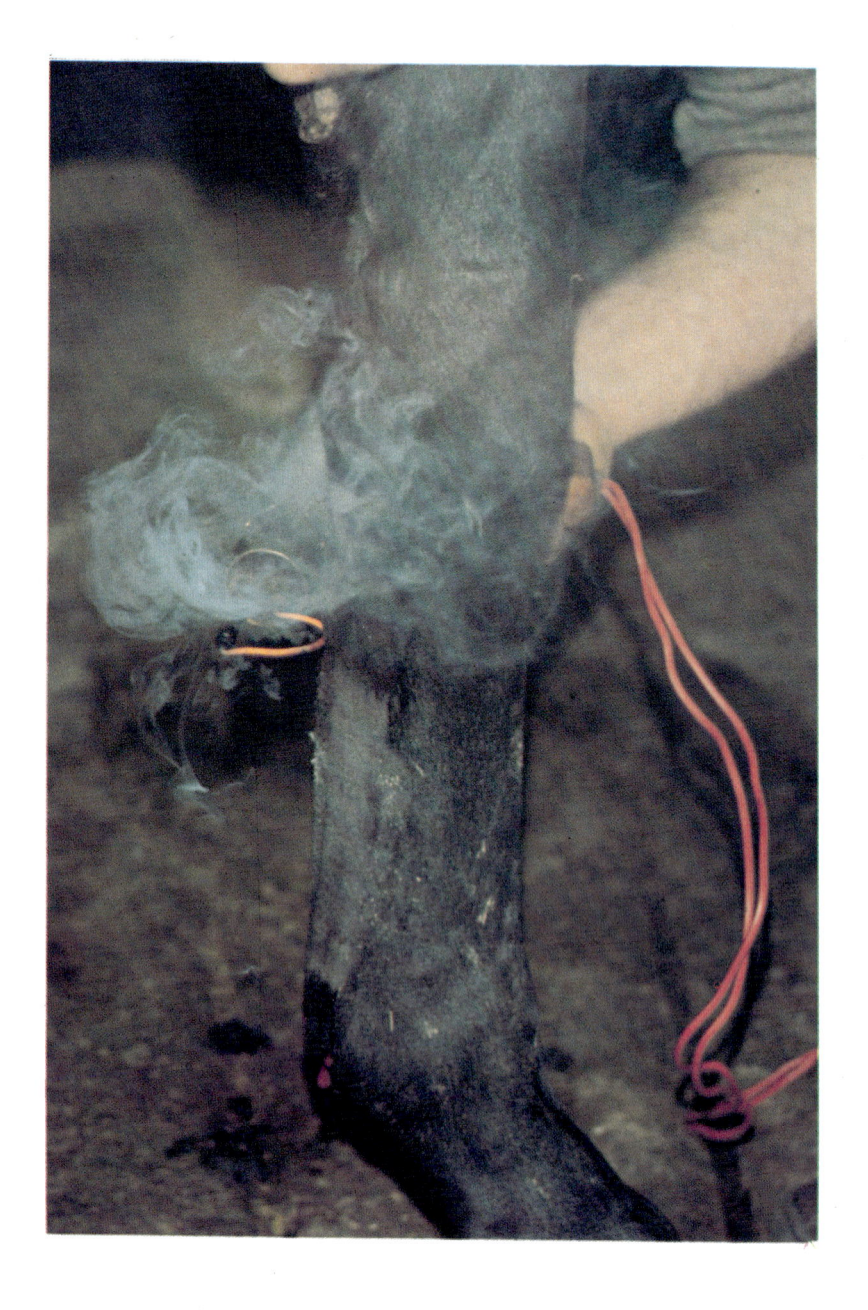

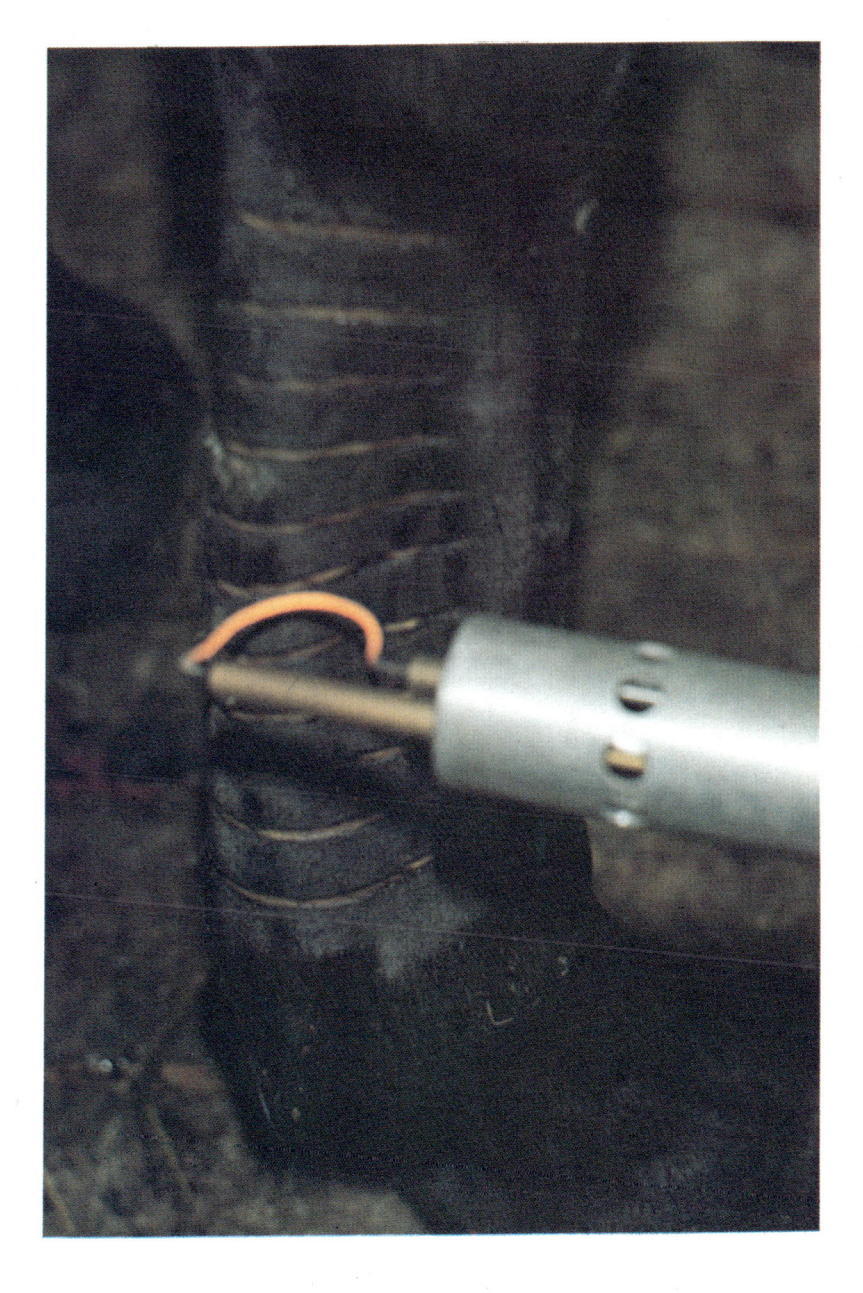

Care must be taken not to burn the skin too deeply, thereby penetrating to the underlying structures.

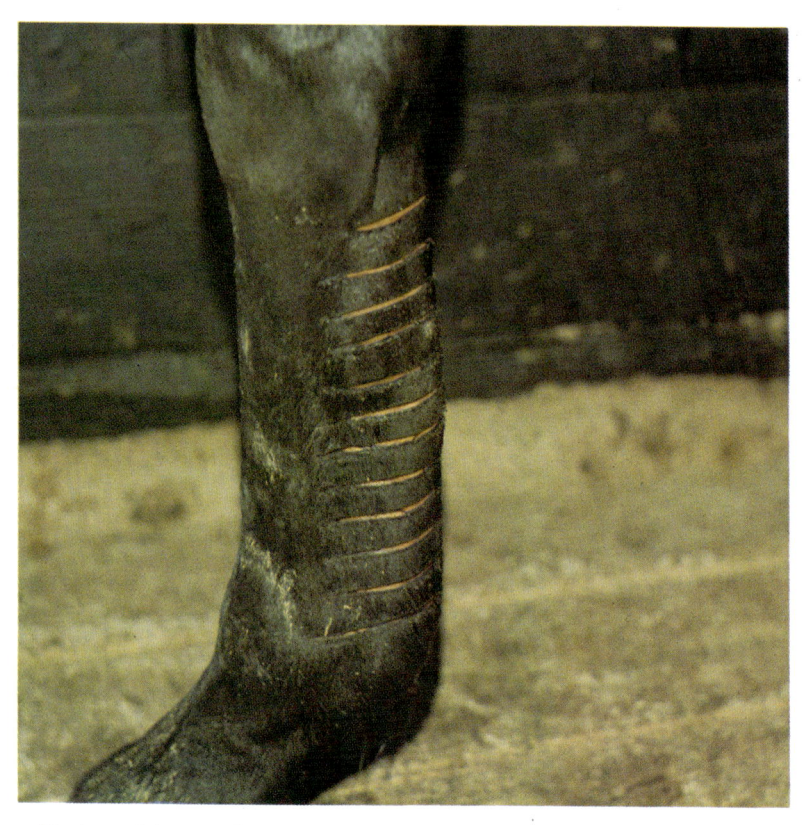

Horizontal lines are burnt into the skin about ½ to ¾ inch (2 cm) apart. The horse is given a local anaesthetic and remains standing.

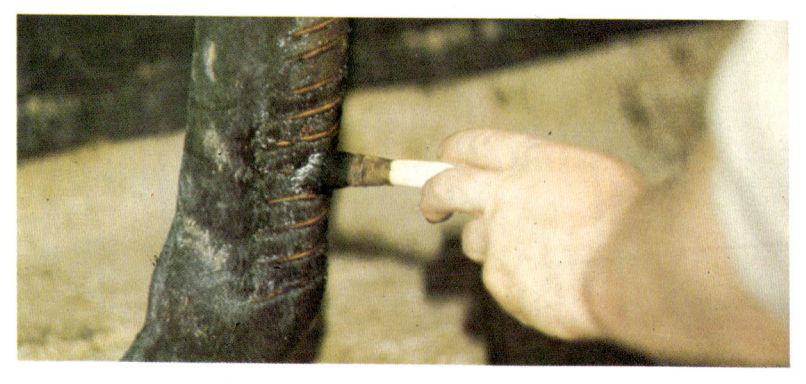

After line firing the wounds are painted with a solution of 25 per cent Glycarin, 25 per cent Phenol and 50 per cent Iodine solution, which acts as a healing agent and counter-irritant to the burnt tissue.

saddle still loosely on his back. Let him cool down quietly before food or water is given or he is turned out. This practice will save many a cough or cold developing, as well as preventing sore backs (see Chapter 4).

Breaking out

On returning from exercise some horses appear to cool off to a normal temperature and then for no apparent reason 'break out' and sweat over large areas of the body. It is often most noticeable on the chest and beween the forelegs as an area of cold, sticky wetness which will not dry off. This is mostly caused by fatigue and the over-exertion of an unfit or partly fit animal. It will also be found to occur more often in animals with a nervous disposition or those having a large proportion of 'blood' in their breeding. To understand why this breaking out takes place it is necessary to understand the basic functions of the skin and sweat glands and how they work.

The skin is responsible for three important functions: protecting underlying tissue and preventing germs from getting in; enabling the body to expel waste products through the sweat glands; regulating the body's temperature at about 101°F (37°–38°C) by adjusting the sweat glands to shut down in order to conserve heat or to open up to release body fluid and with it unwanted heat in an effort to reduce the body temperature. The sweat glands in the skin are controlled and regulated by the central nervous system, which originates in the brain and controls every movement and function of the body. As a result of general fatigue the normal activities of the nervous system are impaired, and its action on the muscles that control the opening and closing of the pores of the skin is weakened. As a result the pores remain partly open, allowing continued sweating to take place. In bad cases no amount of drying off will suffice and the sweating will only be stopped when the animal's nervous system recovers and normal tone has returned. In order to assist this and bring about a return to normal functioning as quickly as possible the following treatment should be carried out. Stand the animal in a comfortable, quiet stable allowing plenty of fresh air. Use a rug and possibly leg bandages to help conserve body heat, but do not overdo this, as many layers of heavy clothing will only hamper the body from returning to its normal state. Nourishment must be given in as easily and quickly assimilated form as possible. Glucose is good for this: add about half a pound (500gm) to two-thirds of a bucket of warm water; it will provide instantly available energy as no nervous output is needed to digest it. A bran mash will also work well; after being fed this the animal should be left with some hay. In a couple of hours or even less, things should have returned to normal and no further treatment should be necessary.

Breaking out will sometimes occur even though the horse or pony is fit and has done no more work than normal. The condition will be noticed on returning to the stable, and the puzzled owner will be alarmed and at a loss for an explanation. It happens in the autumn, what is known in many parts as 'blackberry time', and is due to the winter coat beginning to form under the skin. It will usually occur on several successive days, perhaps as long as a week or so, if it happens at this time of year it should not give cause for concern.

Broken wind

Apart from simple coughs and colds brought about by neglectful practices leading to minor disturbances of the delicate membranes of the respiratory system, there are also more serious conditions affecting the lungs. These cause coughing that leads to permanent damage, when the horse is loosely referred to as being 'broken winded'. This term is often applied to any condition affecting the normal function of the animal's wind, but is more correctly the name given to the condition left by the breaking down of the connecting tissue of the lungs. To understand what is meant by this and how it comes about one must understand the action that takes place during breathing.

One should imagine the lungs as two large balloons at the end of the breathng tube leading down from the nose and throat, one on either side of the chest, filled with minute air pockets or cells joined together with elastic connecting tissues. As air is inhaled these balloons stretch and fill, and it is here that the oxygen in the air comes into contact with the blood in the many minute cells that make up the interior of the lungs. This sponge-like cellular structure is interwoven with blood vessels so that the blood being pumped through them becomes purified and replenished with oxygen from the inhaled air. The heart muscle is responsible for pumping the blood through the lungs. Blood returning to the heart through the body's circulatory system is pumped from the right ventricle to the lungs, where it becomes oxygenated as described. It then returns to the heart to be pumped through the body by the left ventricle. Behind the lungs (our two sponge-filled balloons) there is a muscular wall called the diaphragm, and behind this is the stomach. (There are, of course, other internal organs but for the purpose of understanding the function of breathing we can ignore these.) In order to begin breathing the body draws back the diaphragm, causing a partial vacuum in front of it; the lungs fill this vacuum by filling with air. This is inhalation. The elastic connecting tissues of the lungs then contract and, together with the reverse action of the diaphragm which is drawn forwards, push out the air. This is exhalation. If any of these minute cells becomes damaged or the connecting tissues break, the many small cellular air pockets will be torn apart, and a smaller number of larger ones will be formed. This will cause loss of elasticity to the whole lung, and also lessens the surface area that can be provided for the blood to become oxygenated. To make up for this loss of air intake and the accompanying loss of blood purification, the rate of breathing will have to be increased. Breathing becomes more laboured, and to make up for the loss of elastic recoil of the damaged lungs the stomach muscles will be brought into use to assist the diaphragm in exhalation. When this happens the internal organs situated behind the diaphragm are forced against it, further reducing lung capacity. It will readily be appreciated that a vicious circle is set up with the onset of exercise, and the stress from this produces the cough or broken wind.

The condition is more likely to occur in heavier horses and cobs that are barrel-chested, but can happen with all types of horses. When the condition is non-inflammatory there is no rise in temperature and moderately affected animals may with care continue to work usefully; but there is no cure for the condition once the lungs become damaged, and badly affected animals will only be capable of very light exercise or may be rendered useless. The damage can occur through coughing brought about by the feeding of dusty or mouldy food, by living in an environment polluted with mould or spores or by an animal continuing to work after it has developed a cough. Being subjected to hard work immediately after eating bulky food is also said to be a contributing factor. It follows, therefore, that attention to feeding, so that contaminated foods are avoided, and to stable management, so that only dust- and spore-free bedding is used will go a long way towards preventing this trouble. Do not subject to hard exercise a horse or pony that is soft or coughing, or one that has recently been fed. Any horse that does have broken wind will usually benefit from living out of doors provided its comfort is seen to and all feeds should be damped, including oats and barley, etc.

Coughing can also be the result of irritations and inflammation of the tissue of the throat and breathing passages leading to the lungs, or it can occur as a reflex action produced by a disorder of the stomach. With a cough present because of the latter there should be no significant rise in temperature, nor swollen glands or nasal discharge. It will occur in the stable or when exercising and will be heard as a harsh dry cough. Treatment here is to alleviate the digestive disturbance: violent exercise should be avoided; concentrated foods should be decreased, and mashes and more good hay should be fed. A corrective medicine can also be added to the food and water. If the condition does not clear up within a week then veterinary help should be sought. When some nasal discharge accompanies a cough, together with a slight swelling of the glands beneath the jaw and throat and a little rise in the horse's temperature, then give the animal a laxative diet and make sure it has plenty of fresh air. Clean water should always be available to it, and if you can arrange to take the chill off it in winter so much the better. A medicine or paste for relieving throat irritation can be administered, and exercise should be reduced to walking or leading in hand.

Strangles

This is a contagious bacterial disease, which normally attacks young animals up to the age of five, though in serious outbreaks horses and ponies of all ages can be affected. It is now fairly uncommon, but when outbreaks do occur locally it spreads rapidly. After recovery an animal usually has life-long immunity from further infection. Strangles causes abscess formation in the glands behind and beneath the lower jaw. These glands can be seen to be enlarged, and there is a thick yellow discharge from the nose, accompanied by a cough. At the outset the horse or pony will be dull and not feeding, the nasal discharge will be watery before becoming thick and yellowish, and the head will be held stiffly. The temperature will rise by 3–4°F (2°C) and the throat glands will become hot and swollen. After four or five days the abscess will rupture, and after the resulting discharge a raw sore is left that will leave a scar on healing. Owners should isolate an infected animal, and keep it warm but with plenty of fresh air. If the weather is favourable and shelter is available the horse or pony can be turned out to grass. Patients that are housed should be fed soft food and mashes, preferably by placing the feeding container at floor level, and all utensils should be sterilized. Disinfect all tack and clothing with which the animal has been in contact. The abscess must be encouraged to ripen by the administration of beneficial drugs, which may be obtained by consultation with your vet. After recovery the animal must be given a long rest of at least six weeks to avoid possible damage to the throat that would result in whistling or roaring (see Chapter 6).

Equine influenza

This is a highly infectious condition that usually occurs in local outbreaks about once a year. It is caused by a virus and gives rise to a distressing cough, accompanied by nasal discharge and a depressed condition, with loss of appetite in most cases. The active infection lasts from seven to ten days, but the cough will continue after that, often for weeks if the animal is worked. I believe that many people work affected animals much too soon after infection, thinking that once their horse 'appears' to have got over it then normal work can be resumed. Many horses have been ruined by this haste—they may be left with permanent coughs and lung damage.

In dealing with this disease my advice is, first to take the preventive measure of inoculation. The vet is required to carry this out; the first injection is followed by an initial booster, given four to six weeks later. After this all that is required is an annual jab. Antitetanus can also be included in the one inoculation so the animal is given the benefit of this protection too. The latest veterinary advice tends towards having the influenza inoculations repeated every three months in order to ensure complete protection from the disease, but I regard this as unnecessary. One dose per year will do, and although it may not prevent the horse or pony from becoming infected in a particularly violent outbreak it will ensure that the animal is only slightly affected. In my opinion this is better than trying to protect the animal completely by increased numbers of injections, as if the animal does suffer from a mild attack it can be treated accordingly, whereas if it had been given inoculations every three months, it might appear to have escaped infection entirely during an outbreak, and therefore be subjected to normal work even if it lacks its normal energy. (This lack of energy is due to the presence of the virus in the bloodstream, (but not getting sufficient hold for the horse to show proper symptoms.)

Once symptoms are present the infected horse or pony should be rested and concentrated foods withdrawn. Judgement will be needed to decide—depending on the weather—if it should be lightly blanketed and placed in an airy box, or rugged and turned out to grass for a period. Plenty of fresh water should be available, and mashes fed perhaps with the addition of some boiled barley. If the animal's normal diet contains crushed barley this may continue to be fed if it will eat it during the illness. Good hay will of course be the basis of the diet if the horse or pony is stabled during this period. The period of rest will depend on the severity of the infection, but do not be too much in a hurry to begin normal work again. Three weeks' rest after a mild attack is not too long, and longer than this should be allowed if the animal continues to appear off colour. In any case, wait until a week after recovery seems complete before starting other than very light exercise, and if you find that the animal coughs when exerted cut down the work until the coughing stops. On no account risk permanent damage to your horse or pony's lungs by working it too soon, no matter how great the temptation.

In cases where no inoculation has been carried out and where there is a high rise in temperature and breathing is seen to be quickened one should suspect a serious lung infection, and veterinary help must be summoned. The horse or pony should be rested, kept warm by clothing and left with fresh water. If he can be persuaded to eat a bran mash this can be given, but if it is not eaten up immediately take it away, as it will quickly sour. Failure to observe these precautions, especially if the animal is subjected to exposure or becomes exhausted because of continual exercise, will lower the horse's general resistance; germs that are normally present in the nose and lungs, but are dealt with by a healthy animal, will then become rapidly active and multiply, causing inflammation of the lungs leading to pneumonia. Once this condition develops the animal's temperature will rise by 5 or 6°F (3°C) followed by a quickened breathing and pulse; shivering will occur and the animal will feel cold to the touch. He will look sick and dejected, a thick nasal discharge develops and the slightest exertion will cause coughing. It will be appreciated that before the condition has advanced to this stage the benefit of veterinary treatment is essential; a visit during the first few hours of the onset of these symptoms is most desirable.

6 Whistling and roaring

This is a disease affecting the larynx in which the nerves controlling the vocal chords fail to operate during inhalation and thus leave a partial obstruction to the free passage of air. When inhaling takes place a turbulence is set up producing the whistling or roaring noise. The condition is mostly encountered in larger horses, but occasionally cobs and ponies are affected. Generally it is associated with heavyweight horses and those with long necks, and as it is a hereditary disease, horses and mares so afflicted should not be used for breeding. The sound is only produced as the horse inhales when exercised; when the sound is high-pitched the horse is known as a whistler; if the sound produced is deeper then the term roarer is used. Whistlers are better able to perform work than roarers, but a whistler can develop into a roarer. The respiratory distress produced by roaring can become so great that the animal is prevented from doing even light work.

Anatomy of the larynx

To understand properly what causes whistling and roaring, some knowledge of the working of the larynx is required. The larynx or voice box is situated in the windpipe and consists of vocal chords with cartilages and small muscles. When activated by the nerves, these muscles draw the vocal chords to each side of the throat as the horse inhales, thus leaving the way clear for a free passage of air to be drawn in on its way to the lungs. Imagine the windpipe as a piece of rubber hosepipe about 1½ inches (3 cm) diameter; a section of this about 3 inches (7 cm) long corresponding to the underside of the animal's throat is where the larynx is situated. Looking into the open (nose) end of this tube you would see an elastic chord stretching across the opening from top to bottom on the left-hand side, and another stretching down from top to bottom on the right. Other chords stretch across from the sides of the pipe and are attached to the central chords. When air is pulled down the pipe by the vacuum created by the lungs the chords at either side contract,

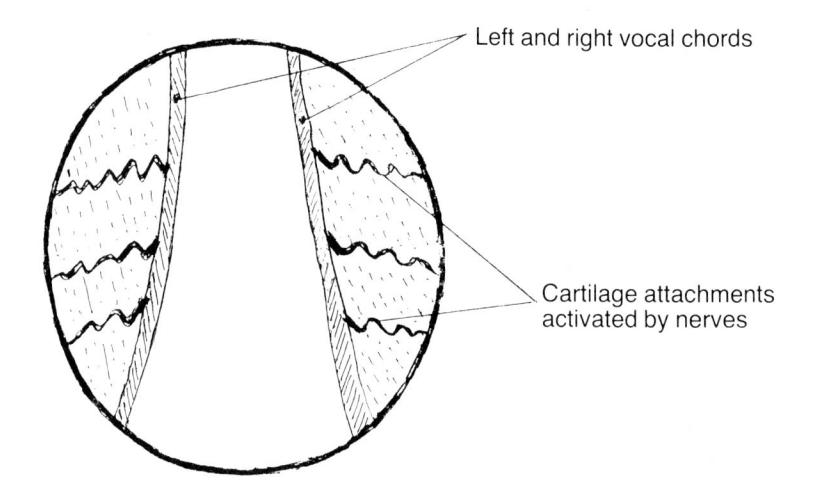

Fig. 3 The larynx: cartilage action to withdraw the vocal chords from the centre of the windpipe on inhalation removes the obstruction to incoming air.

pulling the central chords away from the centre of the pipe to lay flat against the sides. The air being inhaled thus has an unrestricted passage on its way to the lungs. If, however, the side chords (the joining cartilage and muscle) fail to contract the central vocal chords are left suspended in the centre of the pipe, where they partly block the air flow and cause a turbulence. Partial damage to the nerves controlling this action results in a whistle, which leads to a roar when the nerve damage is total and the vocal chords are left slack and vibrating. The shape of the vocal chords and the connecting tissue is such that no muscular action is necessary to enable the free passage of air being exhaled (in the same way that a one-way valve collapses), so no unusual sound is heard.

Causes of damage to the nerves of the larynx

Two main nerves control the small muscles that activate the cartilage to draw aside the vocal chords; one is connected to each side of the larynx. Both these nerves originate in the head and run down the neck with the arteries on either side, but they differ in the route they take to the larynx. The nerve on the right side runs back up the jugular groove to the larynx, the nerve on the left side passes down the chest, round the front leg, then finally winds around the large main artery at the base of the heart before going back to the larynx. The reason for this is not known; it is just a peculiarity of the horse. What is known is that the nerve on the left side, because of its proximity to the large main artery, is subject to damage by the sudden expansion of this artery. If a horse is worked too fast without adequate preparation or is subjected to sudden violent exertion, the abnormal expansion of the large main artery can give rise to partial or complete destruction of the nerve.

Apart from heredity, there are some other causes for this condition, notably the after-effects of strangles. Some say that a septic infection will cause damage to the nerves, and even digestive upsets have been blamed, but this does not explain why only the nerve on the left side is affected. The condition has developed with some horses after coming in from a summer at grass, and a number of experienced owners subscribe to the view that galloping an unfit horse is a contributing factor. Some go as far as to say that because of their extra susceptibility to wind troubles valuable heavyweight horses should not be turned out during periods when flies are bad and they are therefore likely to gallop about. They consider that such horses are best put into a small paddock or kept stabled at such times.

Once whistling or roaring has developed the horse's efficiency will be greatly impaired, and only surgical treatment can effect a possible cure. The operation consists of opening the throat at the point of the larynx and removing part of the membrane lining (ventricle stripping). This causes a surgical injury, which produces fibrous tissues that contract upon healing, pulling with them the offending slack vocal chord to the side of the throat. Once healing is complete the vocal chord stays permanently withdrawn to the side, where it cannot interfere with the airflow. The name of this operation is a Hobday, and it usually gives a fair measure of success.

Other respiratory noises

There are other noises made when breathing that are not caused by nerve damage affecting the larynx. One such noise is that of a horse that is out of condition and has what is known as 'thick wind'. This is a temporary condition, present on both inhaling and exhaling, that disappears as the horse is brought to fitness. Racehorses and eventing types sometimes behave in this way. However, when this noise is present and persists it may be due to a permanent thickening of the mucous membrane of the respiratory system or some other damage to the respiratory passages. A vet's advice should be sought for continuing problems of this kind and it may be that an operation to insert a tube into the windpipe below the larynx will be recommended. Many successful racehorses and eventers have had such a tube fitted, though in the case of the eventer there is some risk to the horse of drowning if falling in deep water.

Noises can also be produced when too severe bitting results in over-flexion. Because of the skeletal angle of the horse's head, there is an obstruction to the free flow of air when the head is flexed in this way. Another noise emitted during expiration by some horses is that produced by the flapping of the false nostril when the animal is fresh at the outset of exercise. It is sometimes referred to as high blowing or trumpeting; it is a habit often associated with high-spirited animals, and is of no consequence.

7 Determining lameness

Lameness is such a vast and complicated subject that full coverage of every aspect and type of lameness affecting horses and ponies is impossible in this book. It requires many years' experience to acquire the skill to diagnose correctly some of the more obscure types of lameness; my task here is to try to give the reader a better understanding of the subject and to make him or her aware of some of the more usual causes.

The name lameness is given when the occurrence of pain, weakness, injury or disease causes continuous or intermittent interruption to the normal working of a horse or pony's limbs. It may only be noticeable at certain speeds or paces and may diminish or increase with exercise. The conformation of the animal is often associated with problems leading to lameness, and the severity of the work that the horse or pony is expected to perform will also play a large part in determining whether lameness is likely to result. There is also the question of the animal being made fit for his task, not being asked to work hard in a 'soft' condition, and also whether it is up to the weight it is being asked to carry. It is generally supposed that a fit horse will carry one-sixth to one-fifth of its own bodyweight when working strenuously over long distances, and that 20–25 lbs (10 kg) can be added to that if the work is to be slower and of shorter duration. In my opinion much will depend upon the conformation of the horse if working to the upper limits of this scale; its legs, and the muscles of the back and loins, would have to be in the peak of condition. If this were not so then lameness, caused by tiredness leading to injury, would be likely to result.

It is true to say that when lameness does occur the seat of the trouble is most commonly found below the knee in the forelegs and in the hind legs from the hock down. Lameness from the shoulder, hip or stifle does occur, but the incidence of trouble from these sites is much less common. I would also suggest that when the seat of the problem is very difficult to determine the foot should always be regarded with suspicion. If the trouble is finally proved to originate

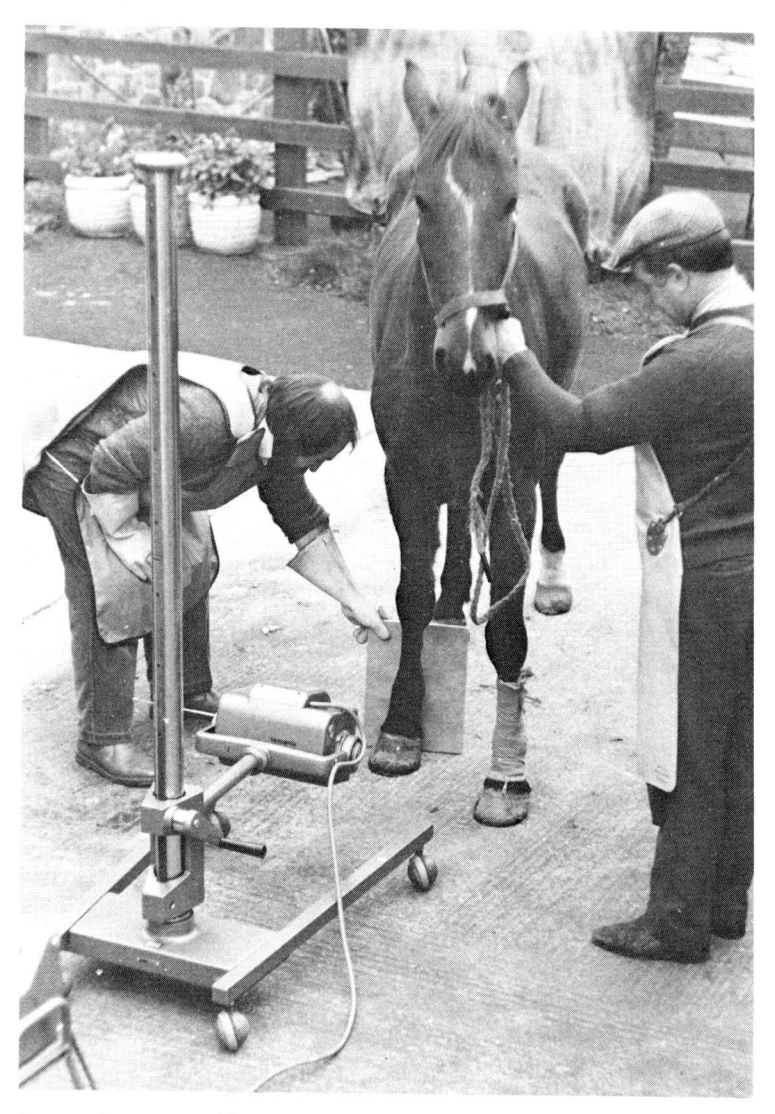

A vet using a portable X-ray unit to confirm a suspected fracture of the pastern.

in the foot, judging exactly what has gone wrong within the complicated structure of the hoof can nowadays be readily ascertained by the vet using x-ray equipment.

If in doubt about the location of lameness in a particular limb, vets today will use a technique known as 'nerve-blocking'. An injection is given at certain points of the lower leg to deaden the

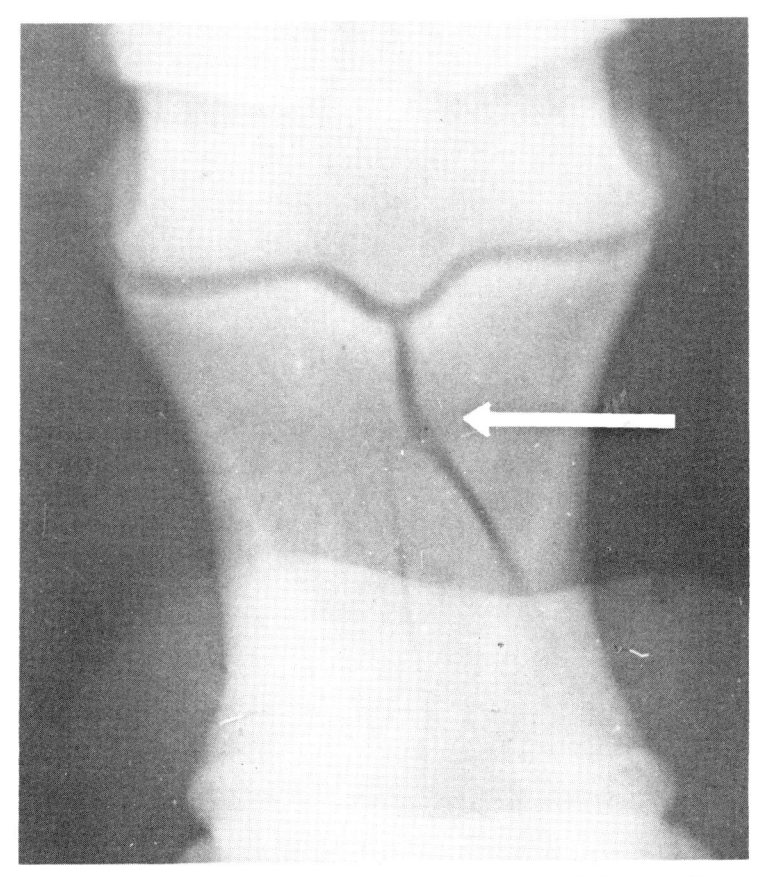

X-ray picture showing a fracture of the pastern. Although the main fracture extends to the fetlock joint above, it was successfully pinned and the pony completely recovered.

nerves and the horse or pony is then trotted up to see if the lameness persists. For example, if a horse or pony has been diagnosed as lame in its off foreleg but there are no visible signs of trouble, the vet will often 'block' the nerve controlling the leg below the fetlock. If the animal then goes sound it can be concluded that the trouble is in the lower extremities below the point that has been deadened. Of course, in many cases the cause and location of lameness will be made apparent by swelling, heat or other signs, and appropriate action can be taken without the need for expensive diagnostic techniques.

If you need to examine a horse or pony for front-end lameness you must choose a hard and level surface and then have someone trot the animal towards you. It should not be ridden for this but led by the helper with a rope attached to the headcollar, and the animal must be given plenty of freedom to move its head and neck. It is important to use a headcollar and rope for this purpose and not to trot the horse or pony up in a bit and bridle; when led from the bit many horses and ponies will not run evenly and will often hang their heads to one side, giving the appearance of being lame in front. The test is best done by taking the horse or pony directly from its stable after it has been resting, as this is when stiffness will be most

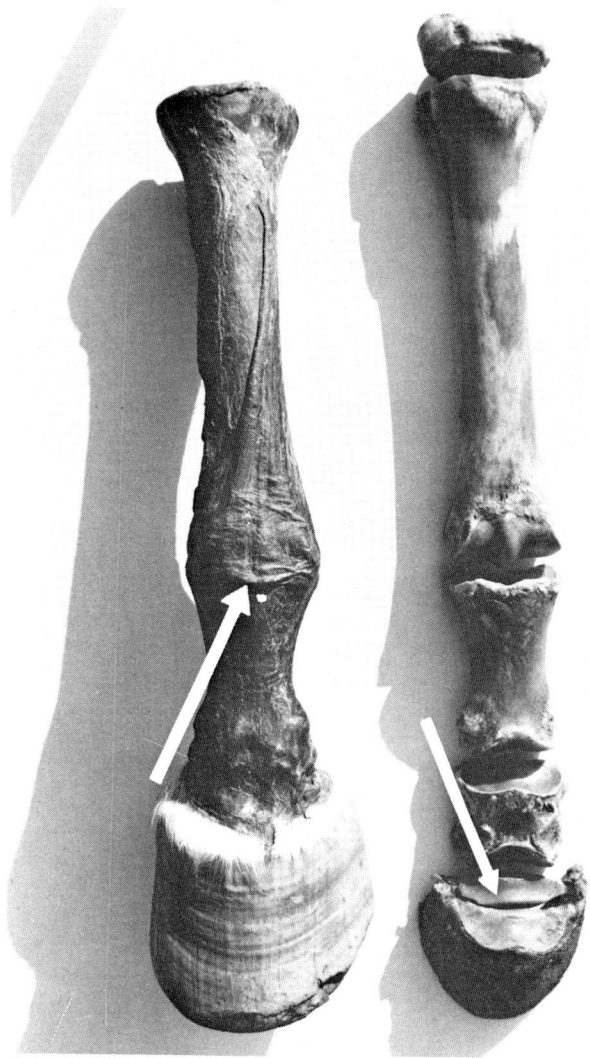

Front view of the bones of the foreleg: the fetlock joint is arrowed left and the navicular bone arrowed right.

pronounced. With lameness in the foreleg the animal will often be seen to 'favour' the affected leg by dropping more heavily on the sound one as it strikes the ground. Standing directly in front as it is trotted up to you, watch its head and if the level of the poll is seen to rise and fall with each step this is a good indication that foreleg lameness is present. The lame leg is the one that comes into contact with the ground as the poll rises.

To determine hind leg lameness the animal should be trotted away from you. Standing directly behind it you must concentrate on whether the movement of the hindquarters is level. If one side of the quarters appears to 'dip' as the horse or pony trots away from you it usually indicates hind leg lameness on the opposite side to that on which the dipping occurs. This diagnosis, is not however infallible, as in many cases of hock lameness the animal will take a shortened stride with the affected leg and drag the toe with the hock not being fully flexed; this will cause the quarters on the affected side to stay lower with each step. In such cases the toe of the shoe of the affected leg often becomes unduly worn down.

With a lameness originating in the shoulder, the foreleg on the affected side will also be seen to take a shorter step than the sound one and the toe will be seen to drag. A horse or pony with a shoulder injury will sometimes 'rest' the foreleg on the affected side by placing the weight on the toe only, whereas if there is trouble in the foot it will often 'point' the affected leg by standing with it placed fully on the ground but farther forward than normal. If a horse or pony is seen continually to lift up a front foot when standing at rest this is a sure sign of pain and and trouble affecting that foot. A sound animal at rest should stand evenly on its front feet with its weight distributed equally. However, many horses and ponies will rest a hind foot by placing it forwards on its toe, and this occurrence in the hind feet should be disregarded. Another useful sign that can indicate pain in the hock joint, such as when a spavin is present, is when an animal only comes up onto the toe of one back leg when staling instead of rising up onto both hind toes. Leaving one hind foot in contact with the ground and balancing its weight on the toe of the other leg when staling can indicate an unwillingness to place an added strain on the joint of the unlifted leg. It is not an infallible sign, however, as most horses tend to be one-sided and to favour one side (as do left- and right-handed people) but it is worth considering.

It will be appreciated that determining the seat of lameness is no easy matter, and unless the cause is obvious it takes a skilful person with years of experience to determine and locate the trouble. It is true to say that many a sound limb has been treated on a lame animal instead of the unsound one! Another thing to remember is that an animal that is lame in both forelegs will often appear sound

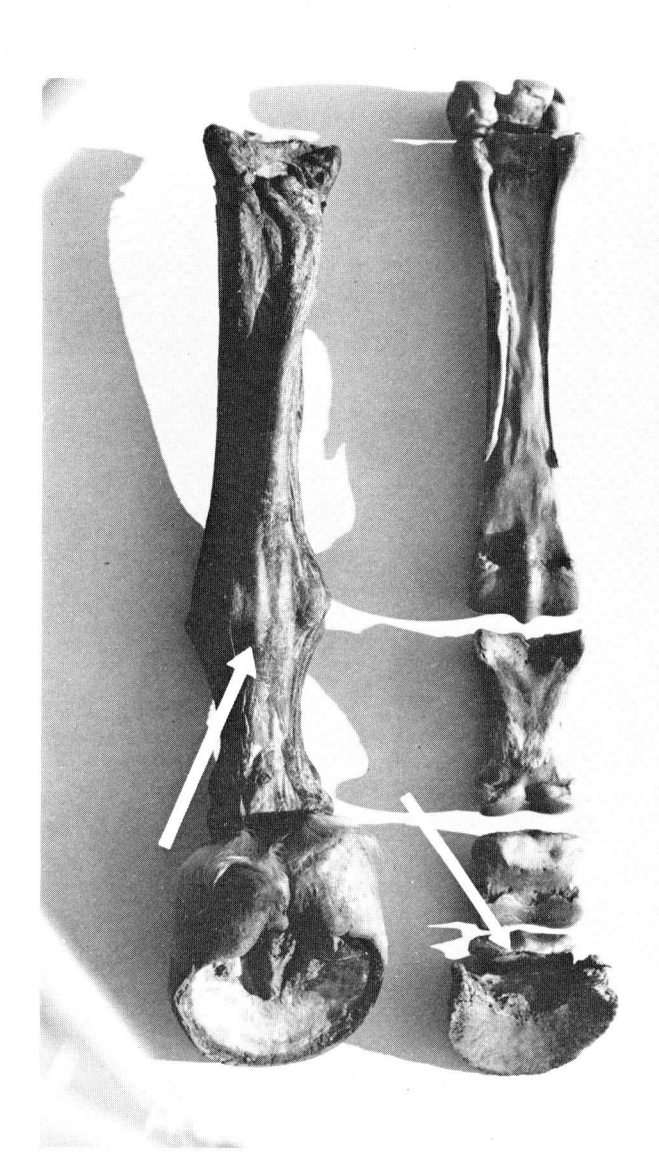

Rear view of the bones of the foreleg: the fetlock joint is arrowed left and the navicular bone arrowed right. Note also the splint bones running down each side of the cannon bone (top right).

except for stepping short or having a pottering stride. For this reason many unscrupulous horse dealers of bygone days would 'hammer' the sound front foot of a lame horse or pony and then sell it as sound to an unsuspecting buyer.

To sum up the procedure for ascertaining lameness, I would say, first, always concentrate your examination on either forelegs or hind legs, but never try to watch both at the same time. Second, if there is no history to lead you to suspect a particular cause then when you have established which end is the lame end, look for the trouble below the knee in front or from the hock downwards at the rear. Always start from the bottom and work your way up from the foot to the knee or hock. When the seat of lameness is difficult to ascertain it will most often be found in the foot when in front and in the hock when behind.

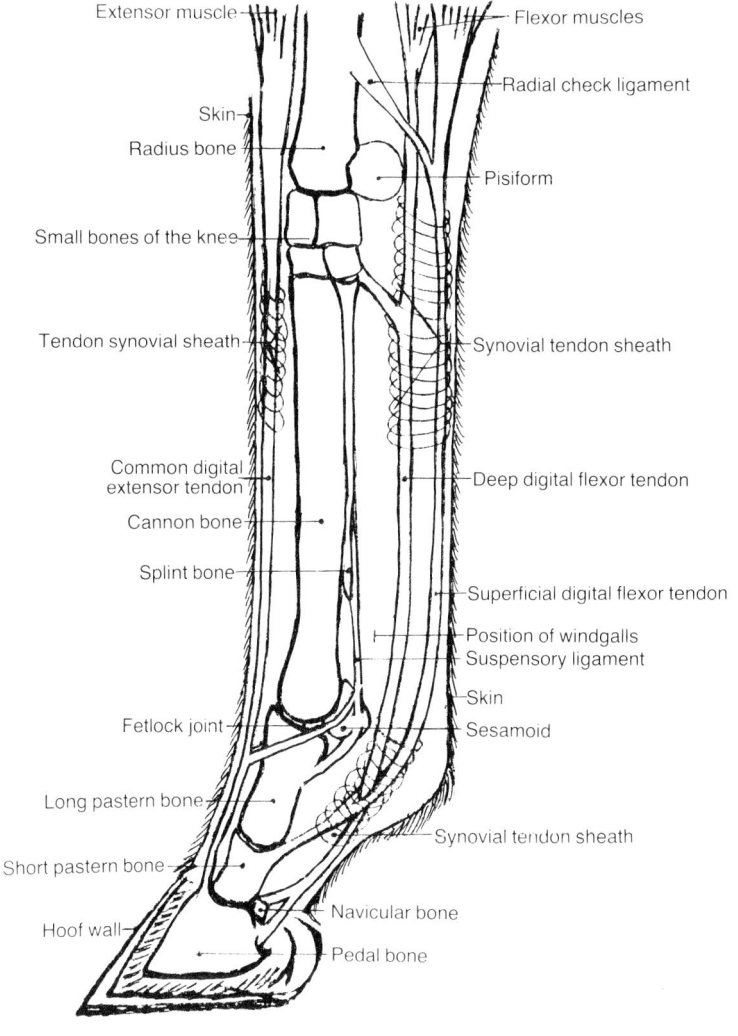

Fig. 4 The mechanics of the foreleg.

8 Troubles affecting the lower leg

Tendon and ligament sprains

In order to understand what happens to cause a tendon sprain or strain it is necessary to have a knowledge of the functions of the horse's leg when locomotion takes place. When the horse propels itself forwards the muscles at the back of the upper leg (in conjunction with the larger masses of body muscles) contract, pulling the flexor tendons upwards. These tendons, situated at the back of the cannon bone, are in fact a continuation of the muscles above but in a different form; they are joined to the bones of the lower leg, including the foot. Ligaments also form attachments both to the tendons and bones and between the bones in order to hold the skeleton together. There are no actual muscles below the knee or the hock.

The leg consists of a series of bones. Those below the knee are all joined together by ligaments, and tendons similar to wires run up the back of the lower leg and join the muscles above. These muscles can be imagined as large rubber straps. When messages from the brain are received via the nerves the muscles will answer by contracting. This muscular contraction causes the tendons to be pulled upwards; they in turn act upon the bones where they join together in the lower leg. As the lowest bone to which the tendons are attached is situated in the foot, which is in contact with the ground and therefore fixed, the resulting pull propels the body forwards. The tendons have only a little elasticity, as they are made up of fibrous strands; this is why they can become sprained or ruptured in the foreleg as they take over from the thrust of the hind legs when a horse is galloping. Muscles, on the other hand, are made up of fleshy fibres that can be activated to contract forcibly; they are softer and possess great elasticity, so they are able to accept violent stretching without spraining. Because of its composition muscle fibre can also be strengthened and enlarged by exercise, though the actual number of fibres composing the muscle does not increase.

Every movement of the body is thus begun by the contraction of

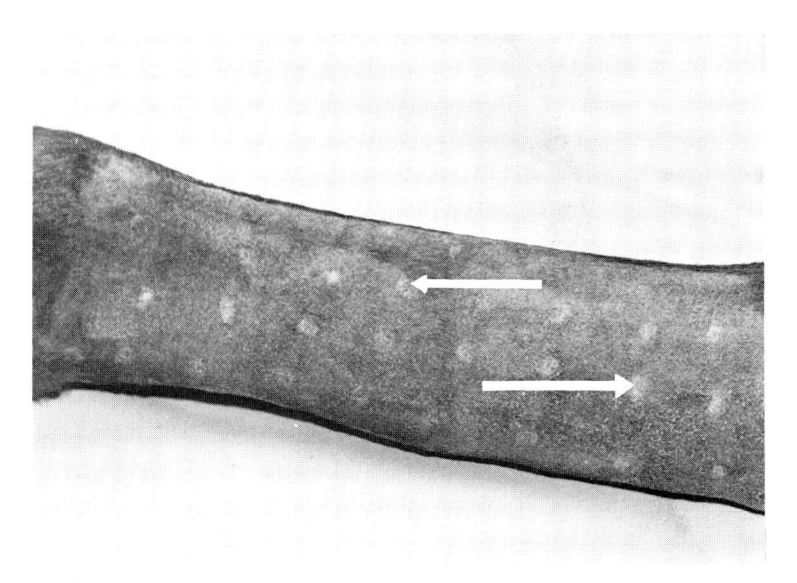

Foreleg with a badly ruptured tendon. The hair of the leg has been shaved off ready for operation. Note the evidence of previous pin firing (spots arrowed).

muscles activating the pulling mechanism just described. At no time is anything 'pushed'. This is because muscles can only forcibly contract and cannot forcibly expand. This may seem hard to believe when watching a horse 'push off' with its hind legs as, for example, when jumping. Nevertheless it is true. The jump is made because the muscles of the buttocks and hind legs contract and, acting upon the back tendons, cause the bones of the hind legs to be pulled violently into a straightened position. The lowest bones situated in the hind feet cannot be forced further downwards because of their contact with the ground, so the body is lifted and propelled upwards to make the jump. Other parts of the body musculature also contribute to this movement, but basically this pulling on the tendons to straighten the limbs is how movement and propulsion is achieved. All the tendons responsible for moving the horse forwards and upwards in this manner are of the 'flexor' variety. Another type of tendon is found at the front of the legs; these are 'extensor' tendons. They work in exactly the same way except that they have the effect of pulling the leg forwards and upwards after it has left the ground, so allowing the leg to touch down again ready for the next propelling movement. Because the action of the extensor tendons and their controlling muscles is much less violent than that of the flexors the extensors are not subject to the same strains or sprains. (They can, of course, suffer from other damage or injury.)

Causes and prevention of sprains

It is the flexor tendons, and particularly those in the foreleg that become strained or sprained, and when this happens it is because of fatigue in the muscles controlling them. This occurs when the muscular action is at its most violent, as when producing the gallop. If a horse is made to gallop when unfit or is forced on when fatigued the muscles are no longer able to use their elasticity to cushion the shock and this is then thrown upon the tendons. As the tendons have very little elasticity they sprain or rupture. This happens chiefly to the flexor tendons of the foreleg, because at the gallop it is the forelegs that carry the burden of propelling the horse or pony, with the hind legs taking up the weight at the end of each forward thrust as the body completes its flight through the air and comes back to the ground. Because the foreleg muscles are working proportionately harder they are more quickly fatigued, so we see the greatest incidence of sprains occurring to the flexor tendons of the forelegs. It follows, therefore, that an important factor in the prevention of sprained tendons is not to gallop an unfit horse or pony or force it on when it is unwilling to do so because of fatigue.

Ligaments are similar to tendons in that they are made of fibres, but they are even less elastic as they have no resilient portion where they connect, such as a tendon has where it joins a muscle. However, ligaments are to some extent protected by the tendons

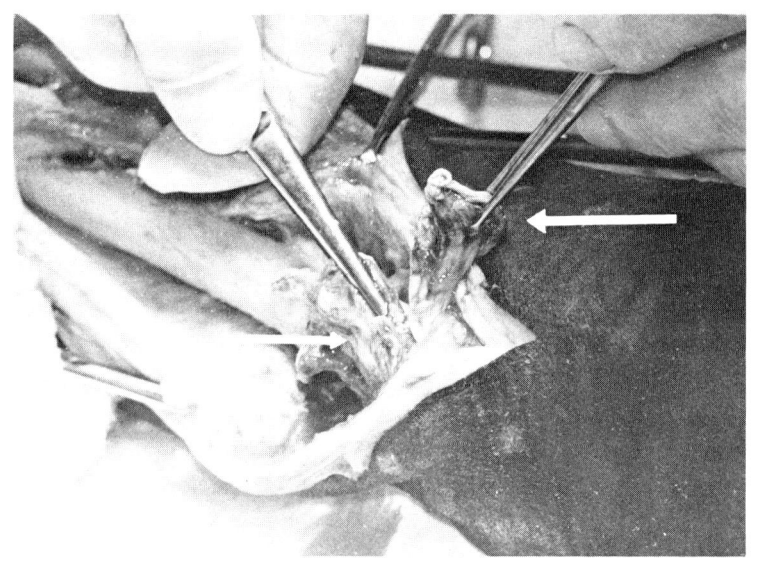

A complete rupture of the superficial digital flexor tendon; the lower broken ends are arrowed. Note the undamaged deep digital flexor tendon that lies beneath.

Surgical repairs being carried out to the severed ends of the s.d.f. tendon.

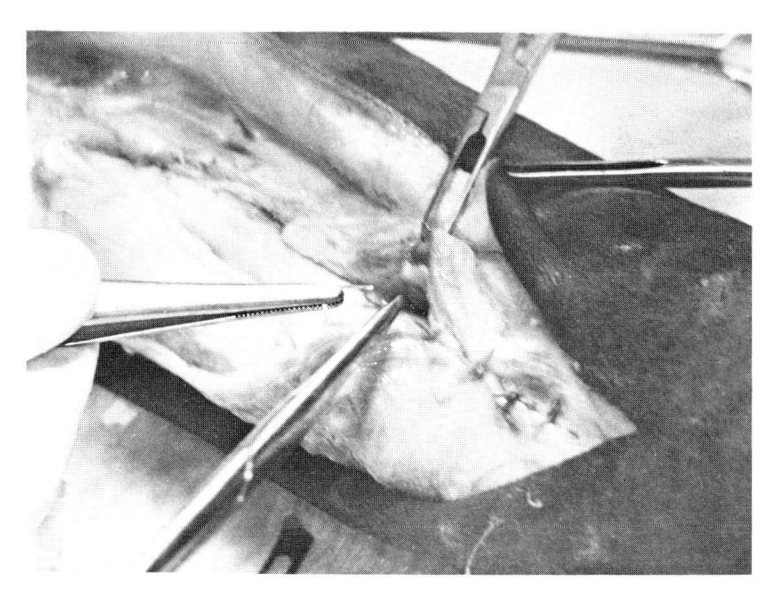

The tendon sheath being drawn together and stitched over the tendon: the outer skin is then stitched. Before the operation the animal is fitted with a special shoe with a raised heel, and is able to stand about one hour after the operation is completed.

from sudden strains unless there is a complete rupture or breakdown of the tendon. A ligament strain is usually brought about more slowly, either by concussion from fast work on hard ground or because of general overwork in an old horse or pony. Ligaments can also be strained when they are called upon to act in an unnatural manner and become wrenched, for example when the legs are carried outwards at an unnatural angle to the body when the horse slips and falls.

Conformation can be an important contributing factor in tendon trouble. Animals that are 'tied in' below the knee, and those with very long pasterns with accentuated slope, will usually be more subject to injury because of these weaknesses in their conformation. It is worth mentioning that the frequency of 'breakdowns' in the lower legs in racehorses is due in no small measure to the fact that these horses, often before they are two years old, are called upon to do work far beyond the capabilities of their immature physique. The lesson here is to bring a horse or pony along slowly until it reaches four or five years old to allow its joints, ligaments and tendons to 'set' before being asked to perform really strenuous work; such an animal will invariably give much better service, with much less trouble from injury, in later life than one that has been made to work hard in its early years. Ponies, too, are much less liable to these troubles.

Treatment

Injuries of the types explained above are serious and will necessitate a long layoff as well as appropriate veterinary treatment. However, not all troubles with the lower leg will be due to sprains or strains to the tendons or ligaments. Some may be only slight injuries causing bruising of the tendon sheath; in such cases a couple of weeks rest is all that is required to effect a cure (see Chapter 4 on bruises). But once a tendon has become 'stretched' beyond its limit, resulting in a sprain, we are really talking about a laceration of the fibres. When this occurs there is acute lameness with a considerable amount of swelling and pain upon pressure. In severe cases there will be a very distinct 'bowing' outwards of the back tendon area. Rupture of these flexor tendons usually occurs midway between the knee and the fetlock joint. When a complete rupture of the superficial or deep flexor tendons has taken place surgical treatment in the form of an operation to join the ruptured tissues will be necessary. This operation to reconnect the broken ends is in many cases quite successful, especially if the rupture has occurred midway between the knee and the fetlock joint so the resulting join does not interfere with the working of these joints. The join does, however, cause a slight shortening of the tendon, and the animal will invariably not work again with quite the same speed or stride. (A racehorse, for instance, would never be likely to race again successfully after this operation, though it would be able to perform less demanding work.) In addition, the cellular structure of the healed tissue does not conform to its original 'in line' formation but becomes cobbled together, losing some of the strength of normal undamaged tissue. To overcome this problem a technique has been developed using a carbon fibre rod, which is implanted in the damaged tendon and induces the cellular growth to follow a straight line during the healing process; this is said to give added strength to the join. This does have the effect of slightly shortening the tendon, so both forelegs have the carbon fibre implanted to ensure that the horse subsequently has an even stride.

Luckily, in many cases of lameness due to tendon trouble the damage is not so severe that operations of this kind are necessary. Early treatment for sprains of the tendons should be carried out by covering the affected part with cotton wool or other wadding spread evenly over the area and then applying pressure to this with a bandage. Crêpe bandage is best for this, and it should be put on tightly enough to give good supporting pressure over the whole area. It must be removed night and morning for the leg to be examined for any uneven denting or 'cutting' and should then immediately be replaced. If it is found to have become looser after the first few hours this is a good sign and shows it is having the desired effect. Bandaging should be continued for a week to ten days; after the first day or two, when the pain has worn off, massage the area by hand: rubbing will help to promote circulation and the removal of unwanted fluid. This should be carried out night and morning when the bandages are removed for inspection of the limb. The best method in applying this massage is to lift the foot and support the leg while carefully rubbing the affected part with your other hand. The joints should also be gently bent and flexed at this time without demanding any effort on the part of the animal, and ceasing as soon as any discomfort is noticed. When reapplying the wadding make sure that none of it has become hardened or lumpy, and if any uneven pressure marks are present on the leg wrap a plain cotton cloth round the leg before reapplying the wadding and bandage. If given this treatment early enough before exuded fluid has become established and solidified, even fairly bad sprains will receive marked relief and respond favourably. It is wise to bandage the good leg as well as the injured one so that any undue weight or strain placed upon it by the animal favouring its bad leg will be supported. Both legs will also benefit from the massage when the bandages are removed.

Importance of rest
Rest is of the utmost importance with any injury. In fact it is

probably the greatest single contibuting factor in any healing process, as it allows the body to perform the task of healing itself naturally without the toxic complications brought about by work. The body will always strive to cure itself, and we can only attempt to create the most favourable circumstances for it to do so. This is why rest is essential, and its value should never be underestimated. Even so, gentle exercise after the initial period of complete rest does become necessary. This will ensure that unwanted fluid is pumped away by muscular contraction and relaxation and more active circulation. This is particularly important in the lower leg, where normal circulation relies on the pumping action brought about by movement. Movement will also be desirable to prevent the healing tissue from binding together with tissue which causes adhesions to adjoining structures that will restrict movement in that area. Cold water from a hose pipe played over the area for several minutes half a dozen times a day will help to reduce swelling further by stimulating the blood vessels to contract and limit the exudation of fluid.

After pain and swelling have been alleviated a horse or pony can be turned out if you can be reasonably sure that it will not become excited and gallop about; a small level paddock or enclosed area will be better than a large field. Do not be tempted to return the animal to hard work too soon, and when you do, be sure to bring it up to fitness in very easy stages. After a severe sprain of the lower leg has responded to treatment, three or four months should be allowed to elapse before strenuous galloping is attempted; the first gallops should be brief and take place only on good going.

Blistering

Animals whose legs show a tendency towards tendon sprain are sometimes given a treatment known as blistering. This produces a reaction of the skin in the area around the tendons, causing counter-irritation to relieve a more deep-seated inflammation. Dilation of the blood vessels increases the blood supply and elimination of waste products through the skin is also increased; this is said to speed up the healing processes of the underlying tissues. The blister is produced by the application of a strongly irritating agent, which must not be applied where there is acute inflammation, so this must first be reduced, by the methods already explained. Blistering used to be (and in some places still is) used to 'freshen up' a horse's legs after a hard season of hunting, racing, etc. A properly applied blister leaves the skin thickened, and this is said to act as a natural bandage, giving the tendons more support. The original severe blister has been superseded by a Green non-irritant type, but even so I consider this treatment undesirable. For one thing, because of the pain and discomfort of the blister the animal's defences will increase the supply of fluid to the affected part without a similar increase in circulation, thus producing congestion. If the animal is fed grain or concentrates before the blister is applied it can then be left with a permanent big leg. I also remain unconvinced that by deliberately inflicting injury to the tissues of the skin, there are beneficial changes in the tissues below. I am not alone in this belief, but there is a strong body of opinion that thinks the opposite. Should blistering be decided upon, the delicate skin of the heel must be protected with a liberal dressing of grease or vaseline in case any of the blister runs down the leg, as it can cause unprotected skin in the fold of the heel to crack and become very painful. It is also generally advised to blister both legs, as if only one of them is blistered, the animal will try to place all its weight on the opposite leg. If for some reason this cannot be done the unblistered leg should be bandaged to give it some additional support.

Firing

Firing is a cauterization of the skin that is even more counter-irritant than blistering. This cauterization is carried out using a red-hot iron; two forms of firing, known as 'line firing' and 'pin-firing', are used. The iron used to be repeatedly heated in a fire, but now an electrically heated instrument is used, which produces a red-hot element like a glowing wire for line-firing or a red-hot needle for pin firing.

Line firing is carried out in the area of the back tendons of the forelegs from just below the knee to just above the fetlock, and on the hind legs just inside the hocks in the case of spavin firing and just below the back of the hocks for curb firing. It must be done by a skilled veterinary surgeon, as the horse or pony must first be given a local anaesthetic and great care has to be taken not to burn the skin too deeply. As the name suggests, with line firing a series of horizontal lines are burnt into the skin abut ½ to ¾ inch (2cm) apart. (Placed closer together than this they would be likely to destroy the blood vessels in between, resulting in nothing but dead skin.) The object of line firing is to produce a thickened skin once healing has taken place; this is said to tighten the skin around the tendons, giving them more support.

Pin firing differs from line firing in that the cauterization takes place in the underlying tissues below the skin or into the tendon itself wherever the heated needle penetrates. In some cases it will be used to penetrate down to the bone.

Both types of firing may be given only when any initial inflammation has reduced so that the area to be fired is cool. Veterinary opinion differs about the use of these treatments. Some

A typical 'bowed' tendon indicating tendon sprain; line firing has previously been carried out but the scars are not visible because of the winter coat.

subscribe to the view that only line firing will be beneficial to the relief of tendon troubles and that pin firing is best for bony deposits; others maintain that pin firing is the correct treatment for tendons. I personally subscribe to the former view, but in any case would only recommend either of these treatments as a last resort if the tendons are repeatedly troublesome. I firmly believe that rest alone followed by a period of light exercise without work to allow natural healing to take place is the best treatment. If this does not do the trick I have reservations about whether the use of firing will prove any more

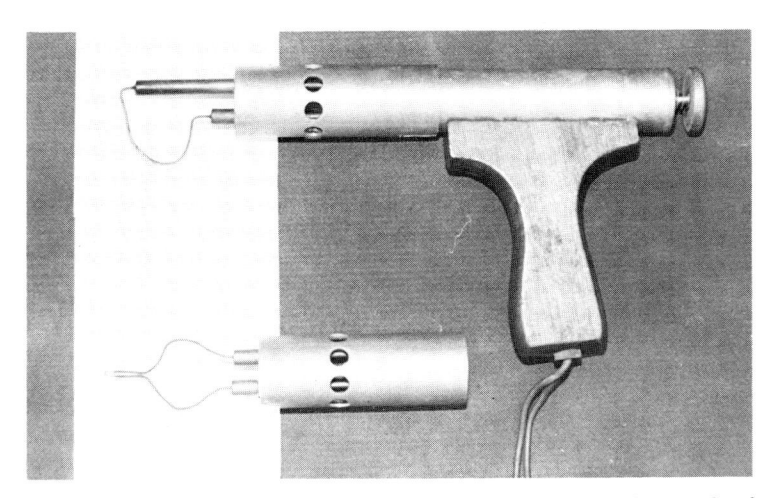

The electrical apparatus used for firing: the attachment for line firing is fitted to the instrument; the pin firing attachment is shown below it.

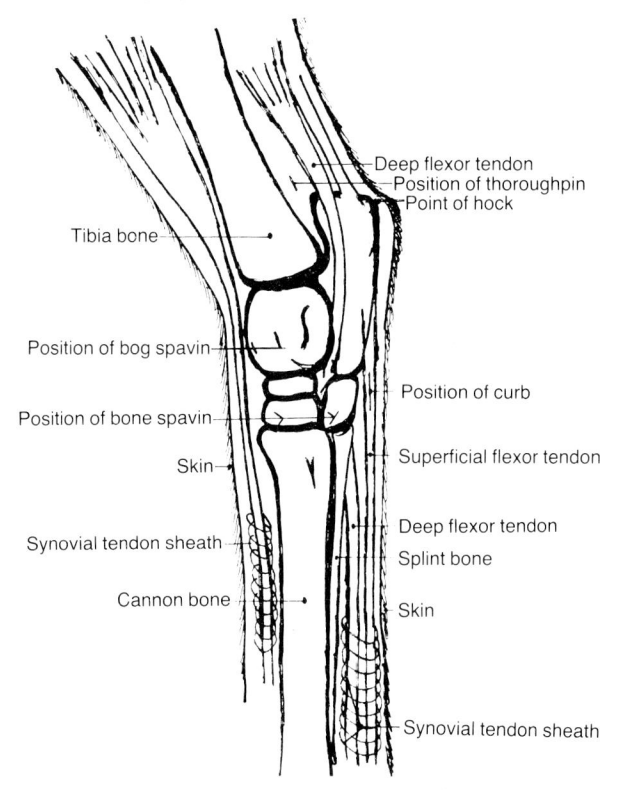

Fig. 5 The inner face of the hock.

successful with an inherent weakness, as I have known many animals to break down again after firing treatment. My conclusion, therefore, is that a predisposition to tendon troubles when an animal is worked hard is due to hereditary weakness, bad conformation or fast work while too young.

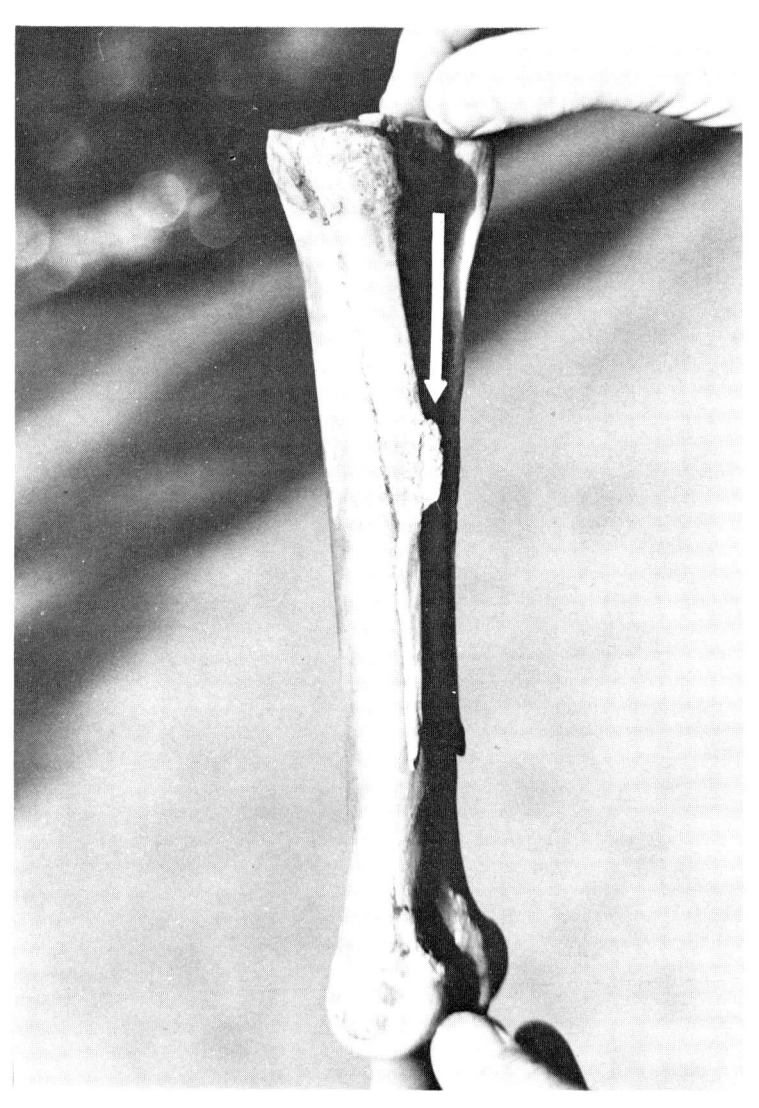

Side view of the splint bone attached to the cannon bone showing a typical splint (arrowed).

Splints

The name 'splint' is used to describe a bony outgrowth found on the splint bone where this connects to the cannon bone. It is usually associated with the foreleg, where the bony deposit will generally be felt on the inside of the leg, but it can also occur with the hind leg,

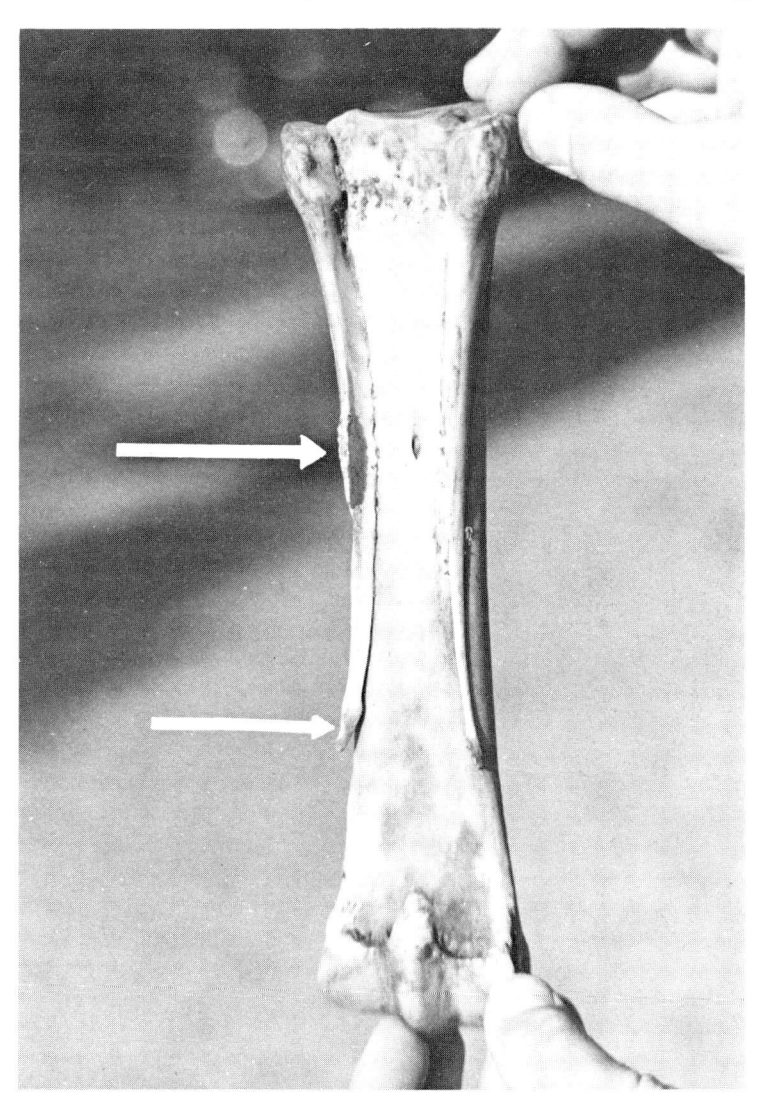

The same splint viewed from the rear (top arrow). The lower arrow shows the nodule at the lower end of the splint bone which remains unattached from the cannon and should not be mistaken for a splint.

where it is generally felt on the outside. The growth will normally be the size of a pea, but splints can be considerably larger. Lameness may or may not be present, depending on the position of the growth on the bone and the age of the animal.

During the initial period of the horse's growth—from foaling to about four years old—the two splint bones that run down the rear of each cannon bone gradually form permanent connections to it. After the age of four the splint bone and the cannon bone become permanently united and no further movement takes place. If, however, an immature horse or pony is given excessive work on hard ground the resulting concussion can cause splints to develop. They may also develop if the bone is struck into from behind when an immature animal is overworked. Faulty shoeing in a young animal can also be a contributing factor by imposing an uneven strain on the inside or outside of the leg.

When they are forming, splints will feel spongy with some heat and will be painful upon pressure. To examine a leg for splints, lift the leg and run your finger and thumb along the groove between the splint bone at the back of the cannon and the suspensory ligament, where you will be able to feel the splint if there is one. When feeling the lower extremity of the splint bones do not mistake the small nodules at the lower end which form a natural 'pip' as splints. If the animal indicates that pressure applied to a suspected splint is painful, make sure you are not pressing too hard: it is easy to pinch the main nerve that runs down the back of the cannon. If the animal is lame then inspect each side of the shoe for uneven wear as this can indicate faulty fitting. It may be, however, that the shoe is correctly fitted to the foot but faulty conformation is causing both uneven wear and the formation of the splint. Rest and reduced work is all that is generally necessary with a young animal for nature to take its course and the bones to become permanently united. With older horses lameness will be unlikely unless the splints form high up, when the action of the knee joint will be affected. In a mature horse or pony splints that feel cool and are not high up will not cause lameness and are of little consequence.

Hock lameness

The hock is a very complicated joint, and can be subject to several, often confusing injuries and diseases. It is generally agreed that diseases affecting the hock are hereditary, so it is unwise to breed from lines known to have a predisposition to such diseases. Many people also think that injuries occur mainly because of bad conformation, but I do not hold entirely to this view as there is also an incidence of injury in hocks that would otherwise be described as good. It must be admitted, however, that hocks of obviously bad conformation, markedly sickle-hocked, cow-hocked, tied-in or small will be likely to impose extra stress on the joint when worked hard. However, it is the severity of the actual work that causes breakdown, as used to be the case among cavalry and draught horses and is now among horses used for strenuous jumping—as in steeplechasing and eventing. Overworking a young horse is also likely to cause trouble.

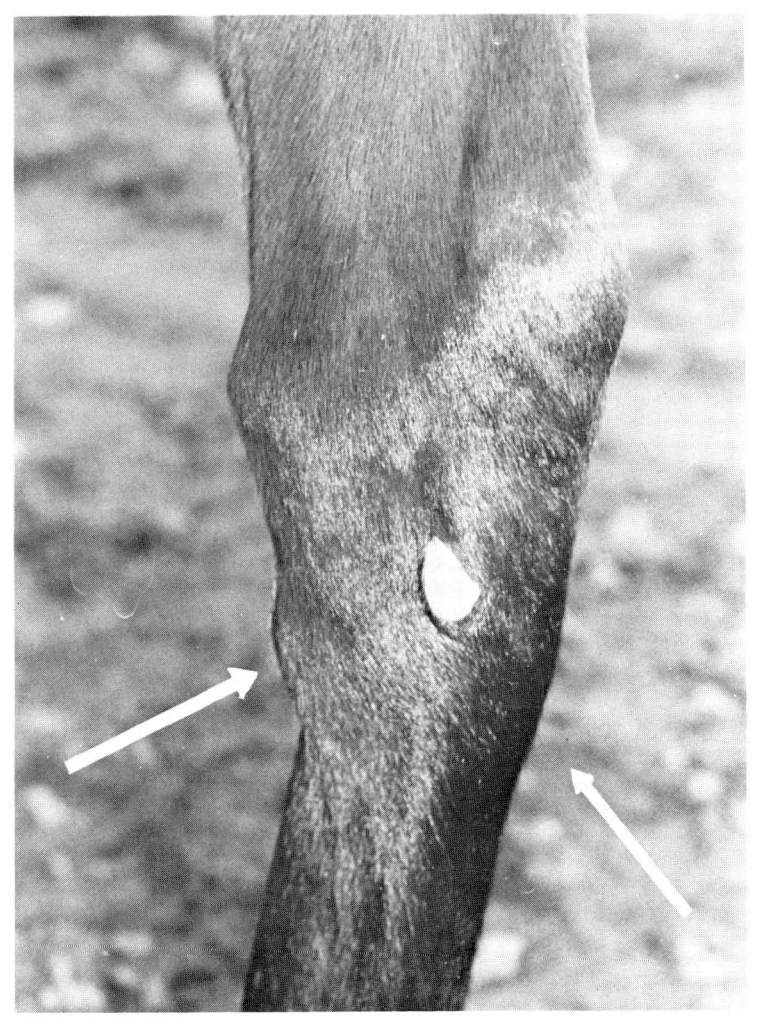

A hock showing a bone spavin (arrowed left); there is also a slight curb present (arrowed right). This condition had been present for three years; there was no heat or pain and the animal was being worked moderatly without lameness.

Bone spavin

This is often referred to simply as a 'spavin' and is a bony enlargement of the lower inner side of the hock joint. It will often give rise to inflammation of the small bones in this area. Lameness is always most noticeable when the animal starts from rest, when it will take a shortened stride with the affected leg and is said to 'step short'. The hock will not be flexed normally and the leg will appear to swing outwards from the stifle with the toe being dragged. A standard test is to hold the leg up tight to the animal's body by the toe for approximately a minute and then have it trotted off smartly immediately the leg is released. When a spavin is present the animal will go very lame for the first few steps. (This will also happen with other injuries or diseases from the hock down.) When a bony adhesion has formed but no inflammation is present there may be no lameness or if there is it will disappear with exercise. When examining the hock joint for a bony enlargement always check both hocks to ascertain any differences in the shape or feel between the two. Rest is necessary for an animal so affected, and in persistent cases pin firing may prove beneficial. In young horses or ponies the lameness will often clear up and disappear, and even up to ten or eleven years of age the outlook is generally favourable. In some animals, however, especially those over the age of twelve, there will be only partial recovery—there will always be some degree of stiffness when the horse begins exercise.

When there is no bony outgrowth but inflammation in the form of an arthritis is present the term 'occult spavin' is used. In this case heat may be felt in the area and the lameness does not diminish with exercise — in some cases it will become more and more pronounced. A long rest of several months will be needed in such cases; line firing or pin firing may help to obtain a partial or complete cure, again depending to a large degree on the animal's age.

Bog spavin

Bog spavin is the enlargement of the synovial bursa, with damage to the membrane lining that controls fluid secretion. It therefore becomes distended, allowing an abnormal accumulation of fluid. It may be caused by a blow, or to animals used for work in which great strain is thrown onto the hocks. Because of their immaturity and greater elasticity of the joint structure, young animals can be particularly affected. There may be heat and lameness; if so and if the condition does not respond to rest, then some experts advise firing. In many instances, however, the swelling is cool, without pain or lameness, and the best treatment is to leave well alone and do nothing. Affected animals often work well throughout life without any interference in the action of the joint: I have even seen the condition in horses that are required to perform gymnastic exercises such as the movements associated with high school performances. With some of the older horses the condition had become an eyesore but they were still able to perform their work.

Thoroughpin

This is a condition similar to bog spavin but not directly connected with it. It occurs at the back of the hock and appears as a lump situated beneath and to either side of the deep flexor tendon just above the point of the hock. When pressed on one side of the tendon it will bulge out on the other side. It is a condition of over-distension of the tendon sheath, and can be caused by galloping through soft ground, rearing or violent kicking. When this happens the swelling will at first be hot and painful, though it will more often be found cool and without pain. Treatment is to apply a cooling astringent when heat is found; the act of rubbing this in will often immediately cause the swelling to subside, and sometimes to disappear completely. Once it is cool thoroughpin rarely interferes with the joint, even if attaining large proportions, so even if the swelling persists it is not of much detriment to the animal and is best left alone.

Lymphangitis

This condition, known as 'big leg', usually occurs in a hind leg. It will be seen as a greatly enlarged leg below the hock, sometimes swelling up to almost twice its normal size. The swelling will at first be hot and tense and can extend right up the leg into the groin. It may be caused by bad feeding management, for example if a horse is fed too much concentrated feed for the work it is doing or is fed normal rations while being rested for a few days; it may also occur as the result of an injury.

Treatment is to cut down all concentrates, not feeding them at all for several days, and to give a laxative feed such as bran mashes. The affected animal must be exercised to stimulate the lymphatic action of the leg, and when lameness disappears should be worked normally. Exercise will almost always bring about a marked reduction in the size of the leg, but the condition has a tendency to recur and eventually to become permanent. A horse so affected will, however, usually work well with little or no detriment to its action.

The entry of bacteria into a wound can cause an infection to the lymphatic system, resulting in a big leg with ulceration; this should always be attended by a vet. Although in mild cases these ulcers will harden and heal in a few days with ordinary simple treatment there is a danger that the deeper subcutaneous tissues may be involved, causing large areas to become infected; this is very difficult to cure and treatment might go on for months.

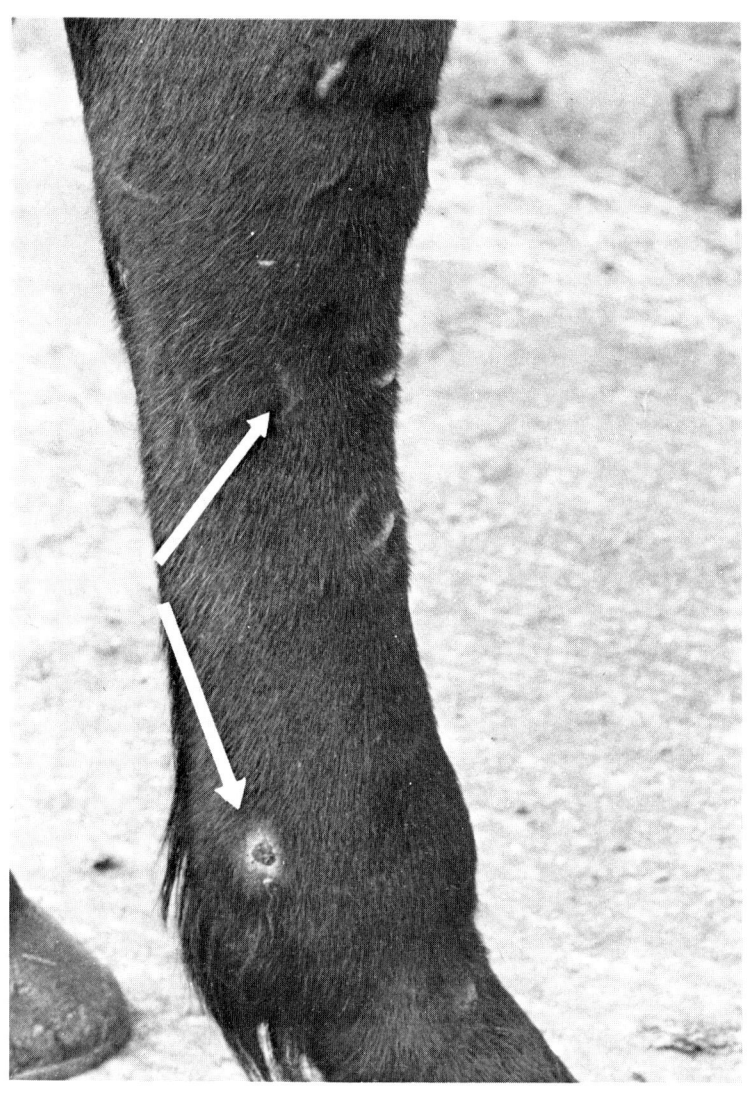

Lymphangitis (commonly called 'big leg') showing swollen and ulcerated lymph glands.

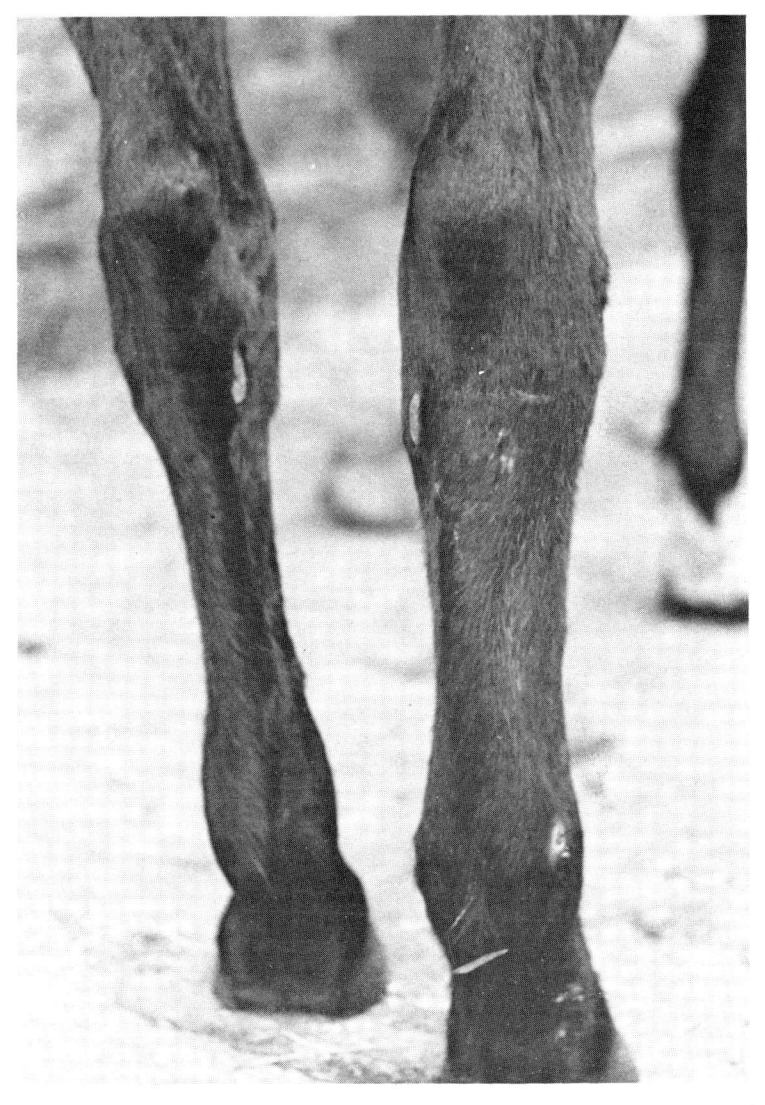

Lymphangitis viewed from the rear: the swelling can be much greater and extend to the groin; it usually reduces with exercise.

The same horse being free-lunged to reduce the swelling. It is sound and moving freely; the off-hind with the lymphangitis can be clearly seen.

Curb

A curb is a sprain to the ligament between the point of hock and the cannon bone below it and occurs at a point about 4 inches (10 cm) below the point of hock. The swelling will extend down the inner side of the hock. It can be seen as an outward curve when viewed from the side, but (in common with other hock injuries) it usually only occurs in one leg at a time so it is best to compare the hocks when trying to distinguish an unusual shape. Causes are slipping after jumping, pulling up sharply with weight thrown on the haunches, and in young horses jumping over obstacles that are too large. It is also said to be hereditary. Lameness can occur immediately after an accident or may not be noticed at all; when there is, it often disappears in a week or two, though the enlargement remains. In bad cases where lameness persists, blistering or firing is sometimes resorted to, but in most cases no lameness will be present and treatment of this kind is unnecesary. In young horses (where its occurrence is most likely) the tendency is for the curb to resolve itself and disappear.

Windgalls

Windgalls are associated with the fetlock joints, and are swellings caused by a distension of the tendon sheaths or synovial membranes

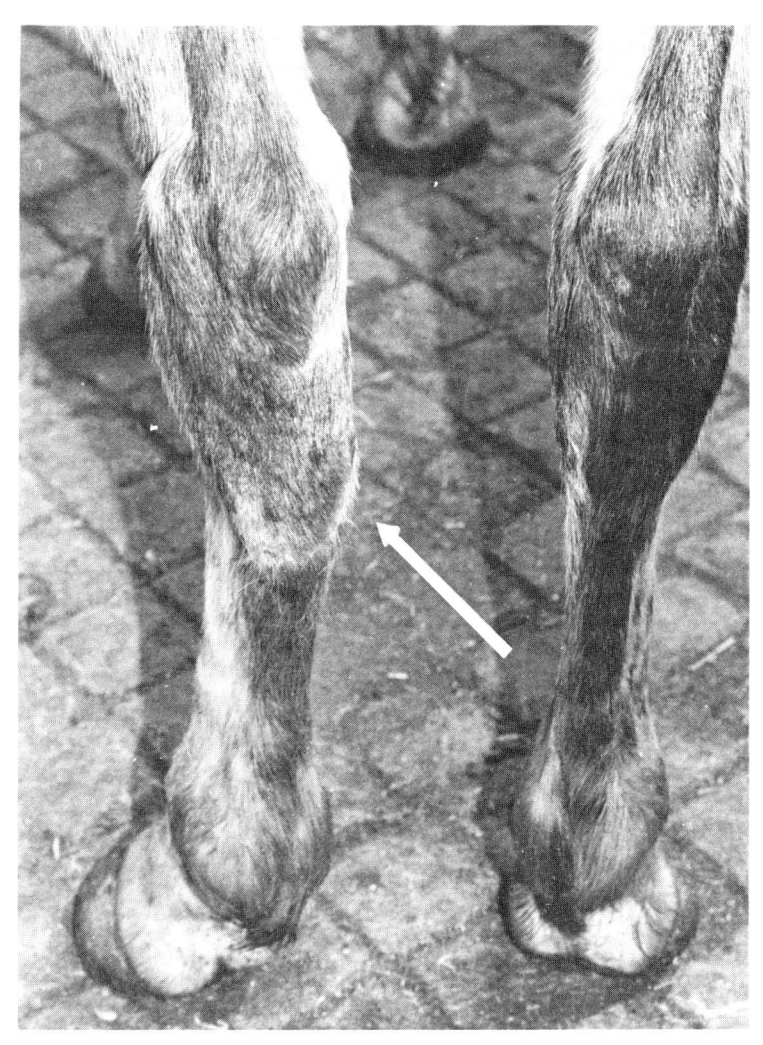

Rear view of a hock showing a curb (arrowed). Evidence of thoroughpin is also present on either side of the point of hock when it is compared to the opposite leg.

present around the joint. They are a sign of work and wear, and many older horses and ponies will be seen to have them. Lameness does not occur in the majority of cases and they are not considered an unsoundness. The swelling will remain constant and feels soft and pliable to the touch. It is only when a windgall becomes bruised by being struck that acute lameness will result, and then it may persist for some time. Generally speaking they are best left alone

85

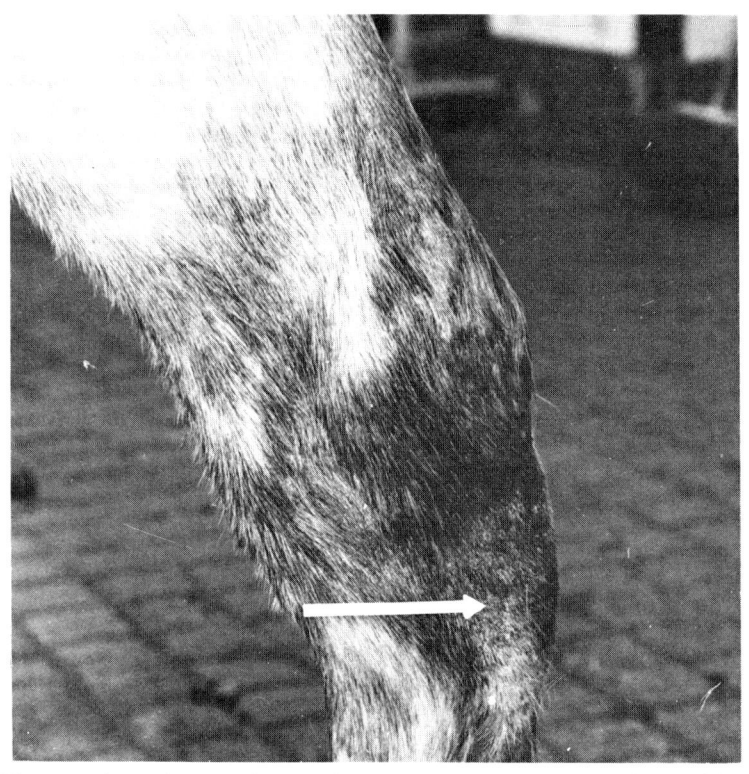

The same leg with the curb viewed from the side (arrowed); the lump of the thoroughpin can also be seen above and to the left of it.

and will remain throughout the animal's life without being a great problem or becoming larger. If they do suffer injury they may be treated by bandaging and cold water applications to reduce the distension.

There are other synovial enlargements, such as capped elbow and capped hock. These are swellings caused by a blow to the point of the joint, which leads to the bursa capsule becoming enlarged with fluid. The exposed position of the hock joint makes this the more likely to be injured by a kick or blow but bad management is a common cause of these injuries: many capped elbows and capped hocks are caused by a horse knocking or scraping the joint on the hard exposed surface of its loose-box when getting up or down. Too thin a layer of bedding will mean that it can easily be scraped away when the animal gets down to rest or have a roll. When the horse gets up again these joints will come into contact with the hard floor, and as the lack of bedding often makes the horse slip and scrabble, the joint bursa can easily receive a blow. It is another reason why I

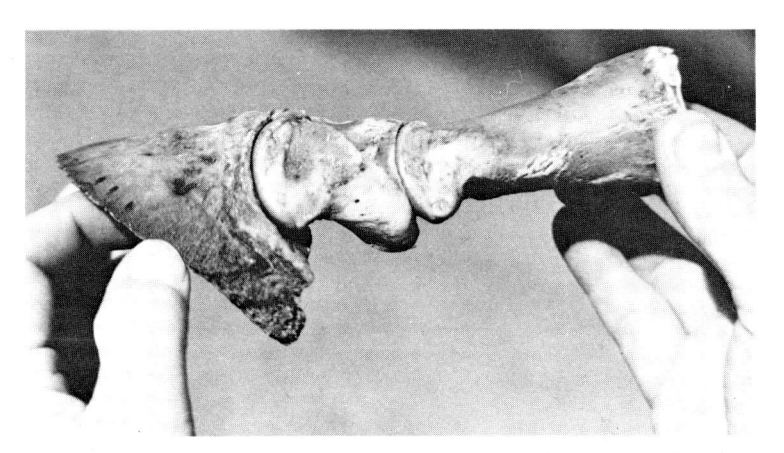

Skeletal structure of the lower leg showing (left to right): the pedal bone, short pastern, long pastern.

Capped elbow

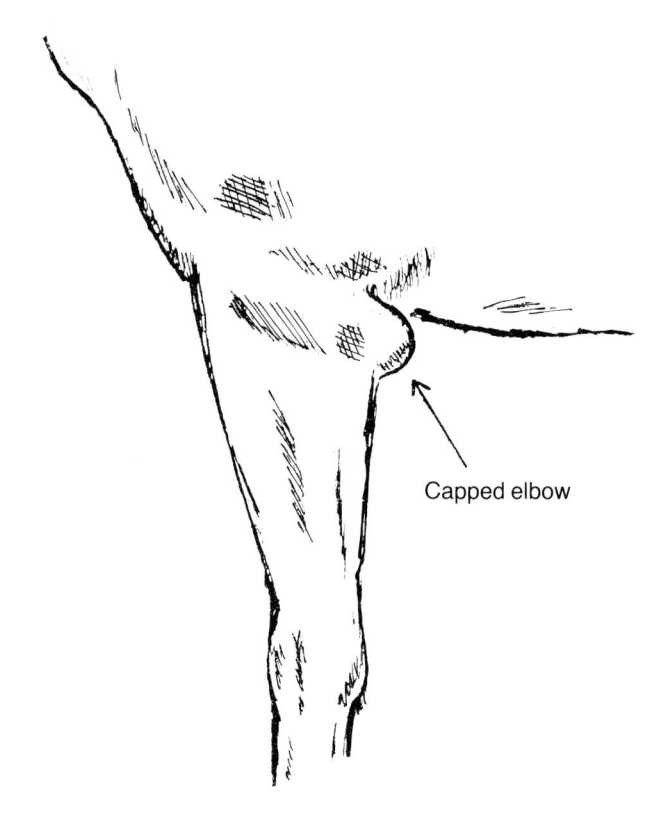

Fig. 6 Capped elbow.

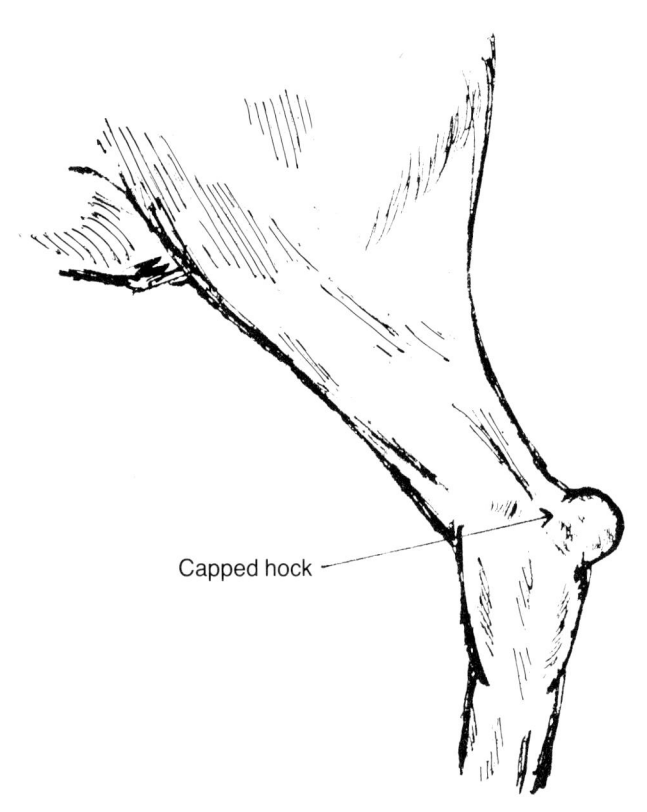

Fig. 7 Capped hock.

prefer deep-littered straw bedding as this cannot happen. Even with good management one can be unlucky and the animal may strike its elbow with its shoe when lying down or getting up, but contact with the hard floor is the most usual cause of this trouble.

The swelling can vary from the size of a golf ball to that of a tennis ball or even larger; if spotted early it can be reduced or kept in check by cold-water applications, astringents and massage. If the swelling will not subside then it may be necessary for the vet to drain off the fluid. The swelling will feel soft and pliable at first, later becoming hard—especially if the tendon sheath is affected—and then lameness may be present. Time and nature usually effect a cure, but although it then rarely causes the animal any trouble it may not completely disappear.

9 Ailments of the foot

Because the foot is such an important part of the horse it deserves a chapter to itself. In order to appreciate some of the things that can go wrong with the foot its structure and working needs to be understood in some detail. Apart from diseases and injuries brought about by the normal working of the horse or pony, bad management and thoughtless practices can give rise to problems with the animal's feet; by avoiding these practices, foot problems can be greatly reduced.

The structure of the foot

The foot is made up of three parts: the bony skeleton, the inner sensitive foot and the outer insensitive foot. The skeletal structure consists of three bones: the short pastern bone extends down to the top of the foot and joins the pedal bone, which is situated within the confines of the foot itself; and a smaller bone, the navicular, is positioned behind and below the joint of the other two, acting as a fulcrum for the flexor tendons. The outer surface of the pedal bone is joined to sensitive laminae ('leaves') which interlace with insensitive laminae that connect to the inner surface of the hard horny wall of the hoof. The interconnecting laminae thus form the bond between the horny hoof wall and the skeleton. The hard, horny outer wall of the hoof is in fact a continuation of the skin, which changes as it continues growing down after reaching the coronary band surrounding the top of the hoof. The change that takes place and the connecting structures are similar to that between human finger and toe nails and the surrounding skin. With the hoof there is also a band of soft, grey-coloured horn, called the periople, which runs just above the coronary band and parallel to it. This expands at the rear of the foot to form the heel; underneath is the rubbery frog, which fits beneath the foot joining into the hard flaky horn of the sole.

The frog has three very important purposes: to cushion the foot (in conjunction with the digital cushion) as it strikes the ground; to

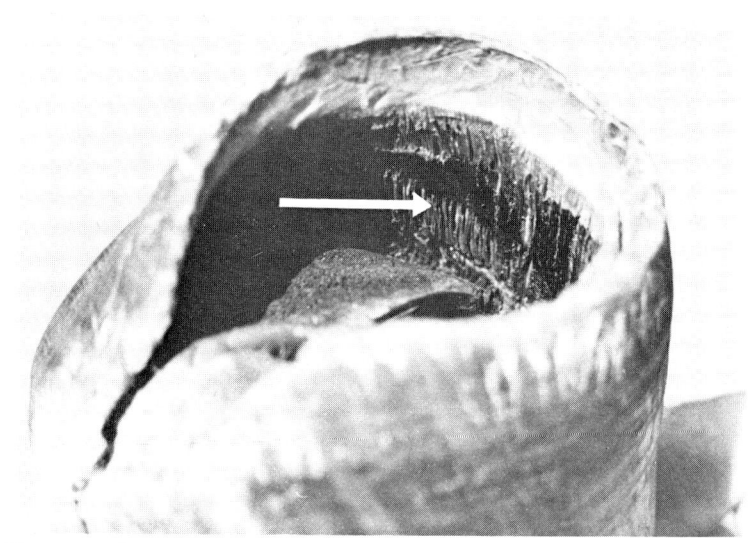

The internal structure of the horny wall of the foot, showing the ridges of insensitive laminae (arrowed) that interlock with the sensitive laminae.

promote grip, because of its wedge-shaped structure and resilience; to act as a pump to increase and speed up circulation to the foot and lower leg. A healthy frog, with good bearing pressure being maintained, is therefore essential. When one thus considers the

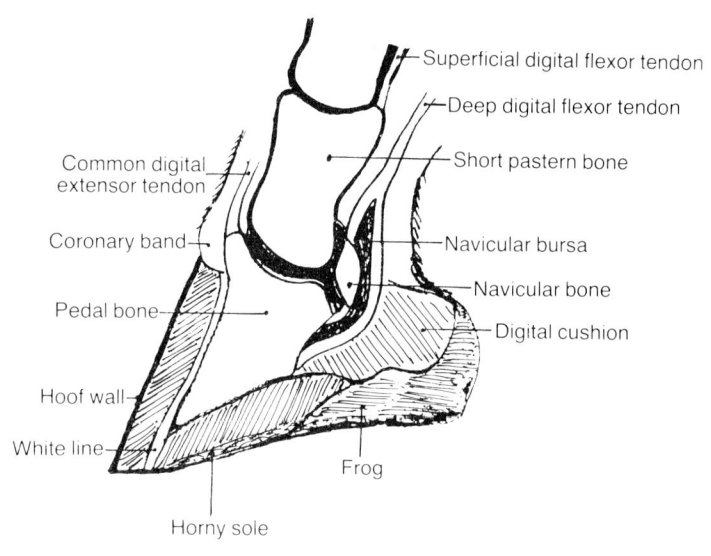

Fig. 8 The structure of the foot.

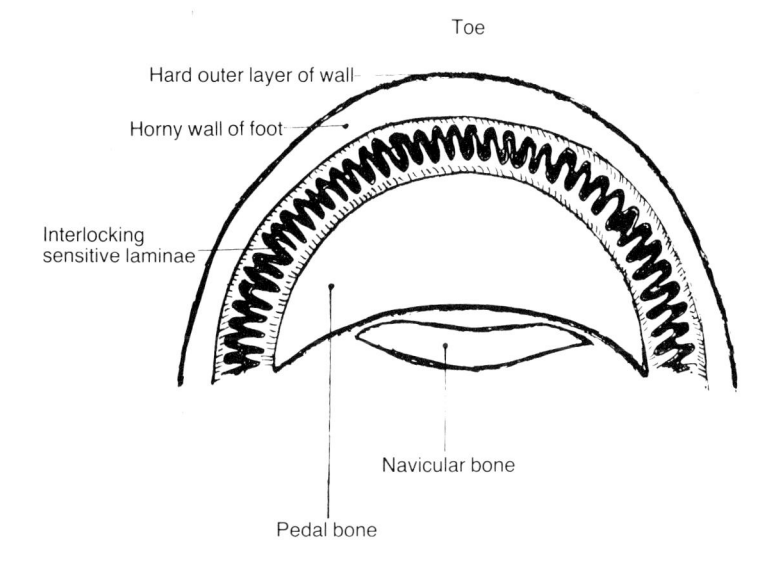

Fig. 9 View of the interior of the hoof from above, showing the joining of the pedal bone to the outer wall of the foot.

intricate structures of the foot and the important role that it plays both in the locomotion of the animal and in bearing the weight not only of the animal but also that of a rider, it is no wonder that when lameness does occur it is customary to look for the seat of the trouble in the foot, where it is very often found. In dealing with foot troubles I will work from the inside outwards, so to speak, by first taking the inner bony skeleton.

Pedal ostitis

The pedal bone is the lowest bone in the foot, and by virtue of that fact takes the weight and the initial concussion resulting from work. If the work is jarring, such as working at fast paces on hard ground, the bone can become bruised and inflamed. Flat, thin-soled animals working over rough, stony going are particularly at risk. In such cases there will usually be some inflammation to the sensitive laminae and connecting tissue as well. The forefeet are the most affected by this trouble; it is rare in the hind feet. When it occurs it will usually be found to affect only one front foot. There will be some heat in the foot, and the horse or pony will step short and will appear tender on the affected foot. The pain will vary in degree, but if there is pain the animal will flinch if the sole is lightly tapped with a hammer. On examination there may also be evidence of a bruised sole. If the condition is noticed early the simplest treatment is to turn the animal out for a prolonged period (about six months) provided this can be on soft ground. If the animal has to be kept stabled then a deep, resilient bed will be essential, as will regular walking exercise to provide circulation to the foot. Good feeding will be necessary, and the addition of cod liver oil to the diet is useful in helping to promote repair.

In all cases of foot trouble, when the vascular system becomes congested the supply of blood is interrupted and the body's repairing functions are prevented from taking place naturally. This difficulty is accentuated because of the hard, inflexible structure of the horny hoof surrounding the foot. The hoof is in fact not totally inflexible, but when it is held by nails to the confines of a shoe it cannot expand very much (though some expansion can take place if the nails are not driven too near to the heels). Because it is difficult to ensure a good supply of blood, light exercise such as walking is needed to help promote circulation. When injury exists this must be on soft going to avoid aggravating the bruised condition. If an animal continues working on hard ground and the congestion is aggravated, the condition can become chronic and the bone will degenerate and become eroded and irregular. Once the disease has advanced to this stage it will be clearly shown by X-ray; the damage incurred is incurable.

Laminitis

Laminitis is another ailment of the foot affecting the laminae and interconnecting tissue between the hoof wall and the pedal bone. Basically what happens in this condition is that the balance of blood flowing into and out of the foot is interrupted. The blood flowing in on the arterial side becomes impeded by insufficient outflow from the venous side, and congestion builds up. Pain results as this pressure increases because it cannot be relieved by enough expansion of the area, which is encased in the hoof wall. The condition is further aggravated because there are no valves in the venous system of the foot to aid in the relief of pressure, and as the feet are at the lowest point of the body any excess fluid will automatically settle there—indeed, the blood returning to the heart has to make a vertical ascent.

An acute attack of laminitis can occur quite suddenly and the horse or pony will become rooted to the spot. All four feet can be affected, but it is more usually the forefeet that suffer, and occasionally only one forefoot. When the forelegs only are affected the hore or pony will try to relieve the weight on them by leaning back on its hind legs. It will usually sweat and blow, generally expressng pain, and there will be a rise in temperature. As a result of the pressure within the hoof, the laminae connecting the pedal bone to the hoof wall 'disconnect' so that instead of being held up in suspension in the foot the pedal bone is released from its position.

Because of this, and owing to the pull exerted on the rear of the bone by the digital flexor tendon, the front end or toe of the bone revolves downwards to press against the sole of the hoof. In severe cases it can actually puncture the sole and protrude through the bottom of the foot. Even when not so severe as this it can still cause the sole to be pushed into a downward, convex position. Indeed, after an acute attack the horse or pony will often be left with a chronic condition that leaves the sole distorted in this way and very sensitive to pressure, with accompanying lameness. Another sign that an animal has had this condition is the presence of laminitic ridges; these are visible on the outer wall of the hoof, and run parallel to each other, merging together at the heel.

As an immediate relieving measure for an acute attack the animal should be stood in very cold water, perhaps by placing the foot in a bucket if only one is affected. If this cannot be done then cold-water poultices or hosing will help. If the animal can be made to walk to relieve the pressure this will indeed help, but in an acute attack will cause the animal great pain. It will, however, be entirely necessary to exercise the animal once the immediate pain and systematic effects have subsided. Frog pressure must be activated in order to pump away the excess fluid. Veterinary treatment will probably take the form of relieving the pressure, either by surgical means or by injections to the vicinity of arteries above the foot to reduce blood supply.

Laminitis can be attributed to many things that can upset the circulation of blood within the feet: excessive work over long distances, causing near exhaustion in an unfit horse or pony not properly conditioned; idleness and want of exercise while being fed a diet well above maintenance needs; a diet too rich in protein introduced suddenly in large measure; injury that results in the horse or pony 'resting' one leg and therefore placing an undue strain on the other. All types of horses and ponies are liable to the condition but ponies, especially those of the hardy type, are most susceptible to one particular form, known as grass laminitis. This is the result of allowing a pony to eat unlimited amounts of lush spring grass, especially when the animal is not being worked hard. It can even occur with some types at periods of the year when grass is not expected to be high in food value, especially when an unseasonal warm spell follows rain and promotes growth. Gluttonous types that gorge themselves will suffer an imbalance through too much rich food, which the system will not be able to cope with, and the result will be the build-up of vascular pressure within the feet. Some ponies will suffer quite severe grass laminitis through this, and the acute effects will be the same as those observed in any other form of the disease. The after-effects, however, are not so serious. Many ponies that have suffered from serious grass laminitis have fully recovered, showing no permanent trace of the disease in their feet. Good management and forethought can play an important part in the prevention of this disease. Feeding routines (as set out in *Horsekeeping—Ownership, Stabling and Feeding*) may be followed to good effect, and ponies should have access to spring grazing restricted by part of the paddock being shut off, or alternatively by their being stabled during part of the day. Care in balanced feeding and only gradual change to different foodstuffs, plus an awareness of the need for progressive exercise to bring an animal to fitness, will all be rewarded by the avoidance of laminitis and other related problems. Cod liver oil added to the feed three times weekly (about one-quarter of a cupful mixed with bran and added to the midday feed) will also help to promote good horn growth as well as helping to put a 'bloom' on the horse or pony's coat.

Navicular disease

This is the disease affecting the navicular bone, which is situated in the hoof behind the pedal bone. It is shaped rather like a shuttle and acts as a fulcrum to the deep flexor tendon that passes over its posterior surface. Its position in the foot makes it somewhat vulnerable to injury from concussion, especially in horses with thin soles, but nature does provide protection from this by providing a bursa or fibrous sac containing fluid that acts as a buffer. When damage does occur, perhaps by a horse treading on or picking up a pointed stone in the hoof, it may be limited to a bruising of the surface tissue of the bone and surrounding structures, in which case resting the animal and giving light exercise on level going will allow the problem to resolve itself. If injury is severe, however, and the damage penetrates to the deeper bone tissue then the problem is much more serious. When the bone becomes damaged erosion can occur to its surface, resulting in the loss of its smooth working surface because of the roughness produced. In some cases lameness will only be slight, in others it will be quite severe but will then disappear or become intermittently slight. The animal may be seen to be uneasy on its forefeet when at rest in the stable and later on will 'point' the affected foot when resting. If both forefeet are affected then each will be placed forwards in turn. This practice will suggest the disease in its early stages and X-rays revealing cavity formation in the bone will confirm the disease, or perhaps only the initial signs of trouble by revealing a shadowy area along the lower margin of the bone. Shadowy areas are not in themselves conclusive, however, as part of the frog can throw shadows on the film and this can be mistaken for erosion.

The navicular bone can occasionally be so severely damaged that it fractures, and when this happens it will usually be in a foot that is shaped wide and open with a shallow-vaulted sole. Damage to the navicular is rarely found in hind feet because they are of a rounder and deeper design. The conclusion is that one wants to see a horse or pony standing evenly on its forefeet, with no fidgeting of either forefoot when it is resting, and neither leg should be pointed. With hind feet the resting of one leg forward onto its toe is normal and not indicative of disease, but with forefeet the practices described should be regarded with the utmost suspicion.

There are several theories about the causes of navicular disease: injury caused by concussion; interference with the normal blood supply; predisposition to the disease from hereditary causes. Some or all of these may be responsible, but positive proof is not conclusive, as many horses worked continuously on hard going (police horses and carriage horses for example) never develop the disease, nor do others that are worked hard on stony going. Also although abnormalities in conformation are passed on and unusual stresses on the joint and navicular bone could be a factor, if it was purely an hereditary disease the number of cases would be legion. It is true, though, that wounds and injury to the area, be they direct or indirect, are a likely cause, and the risk of this is greatly increased by a wide-open, poorly domed sole and a shallow heel. Smaller, bell-shaped feet with a strong vaulted sole and deep heels offer much better protection. It may be for this reason that navicular disease frequently occurs in hunter types and hacks whereas it is not often seen in ponies.

There is no cure for the disease once it has developed. Horses suffering from the disease can be shod with a leather pad to reduce concussion and special shoeing can be carried out but these remedies are only palliative. Neurectomy, called denerving, can be carried out, although many horsemen dislike the thought of this. If the sensory nerves to the affected parts are severed and the operation is carried out by a skilled veterinary surgeon, relief from the constant neuralgia will be given. In many cases the whole foot need not be deprived of all sensation and many horses have been able to work on for many years free from pain and lameness by having this operation. Some have even steeplechased successfully after having such treatment. It must be said, however, that there are dangers when any part of the body is made insensitive to pain because nature intended pain to be the method by which the body is aware that something is wrong. Without the sense of feel or pain other damage such as punctures to the sole, may go unnoticed so detailed inspection to a denerved horse must be regularly and diligently carried out. A riding school horse or any trained and valuable animal could be denerved to extend its useful life, and when used on prepared ground such as that of a riding hall, etc., would face little danger from unseen injury. A denerved horse is technically an unsound horse, and this must be declared when any sale takes place, though I strongly suspect that many people have bought horses unaware that their purchases have been given this operation.

Ringbone

This condition is a bony enlargement to the pastern bones that run down from the fetlock to the foot. When it affects the part covered by a tendon or the joint then lameness will occur. The condition occurs mainly in the forelegs, probably due to the greater strain imposed on these during propulsion and the greater concussion suffered by the forelegs generally. If caused by an injury with inflammation present then rest and normal treatment to remove the inflammation may effect a cure. When a bony outgrowth occurs this will be both felt and seen as an enlargement, on the pastern bone with a high ringbone or as a bulging around the coronet with low ringbone. Hereditary causes, such as those associated with navicular disease, are most often said to be responsible for the development of ringbone, which usually occurs in older horses. The joint affected then becomes permanently stiff, and although the animal may be fairly sound it will have a stiff, stilted action.

Sidebones

Sidebone is a hardening of the cartilage at the bulbs of the heel above the 'wings' of the pedal bone within the hoof itself. When sidebone is present this cartilage, instead of yielding to pressure when pressed in with the fingers as is the normal response, will present a hard resistance. It is not generally associated with riding horses of the lighter types, although any class of horse can have sidebone. Hunter types of heavier build are most at risk, and it is almost always the front feet that are affected. Lameness does not occur as the result of sidebone itself and when there is lameness other causes for it should be sought. Without lameness no treatment of the condition is necessary.

Punctured sole

Injuries to the foot can include wounds inflicted to the sole by sharp objects that penetrate into the sensitive structures; broken bottles or nails are two examples. Infection will then be carried into the deep vascular parts of the foot, and inflammation, pain and pus formation will result. As the pus builds up pressure is created that has no outlet and intense pain is produced. Also swelling will often extend over the area of the back tendons and this swelling, unlike tendon sprain which is 'solid', will dent when pressure is applied with the fingers. Horses and ponies suffering from this type of injury will show clear signs of agony, such as sweating and blowing, and will be unable to bear the pressure of the foot on the ground: even the slightest tap with a hammer will cause great pain. Veterinary treatment to relieve the pressure will be needed: the vet will pare away the sole until it ruptures and the pus is released. This hole should then be made bigger to allow free drainage and hot antiseptic poultices should be applied to relieve the congestion and bring about the complete withdrawal of the pus; anti-tetanus injections should also be given.

Shoeing pricks

Farriers will occasionally place a nail too close to the sensitve structures of the foot, which will cause a pressure, or drive it wrongly so that it pricks the inner sensitive structures. If your horse or pony shows lameness or signs of pain in the foot soon after a visit to the blacksmith, this injury should be suspected. Have the farrier remove the shoe immediately the condition is noticed, and if a nail is found to have penetrated the sensitive structures disinfect the track at once. Prompt action of this kind will often avoid serious trouble. Sometimes the nail will not actually prick the inner sensitive foot but will pass close to it, causing a bulge in the horny wall. This will cause painful pressure to the sensory nerves that can remain even after the nail is withdrawn. Treatment for this is to rest the animal, leaving it unshod, and to relieve the tension by softening the horn by placing the foot in hot water; repeated several times a day for a few days, this treatment will usually result in complete recovery.

A nail embedded in the foot, discovered after exercise. Luckily it was in the thickest part of the bar and did not penetrate to the sensitive structures.

This shoe is not badly worn but has cracked and could lift and become dangerous, proving why regular inspection is necessary.

Corns and bruised sole

A corn is the name given to a bruising of the sole in a particular area, the angle between the wall and the inturned bar of the sole being known as the 'seat of corn'. Corns give rise to quite severe pain and lameness, as they go beyond simple bruising of the sole to affect the underlying sensitive laminae, and in some cases can even go as deep as the bone itself. For this reason they do not lend themselves to remedy as easily as a bruised sole. Pressures causing corns may come from a badly seated shoe, perhaps by its being left on for too long, or from a large stone becoming wedged in the shoe at the seat of corn. Ordinary bruising of the sole can result from exercising over rough, stony ground. If bruising is limited to the sensitive sole and the damage has not gone deeper than this, early relief can be expected. If, on the other hand, the deeper tissues and the pedal bone are affected it may be a very long time before the injured parts become normal. Sometimes a 'pipe corn' will develop, when a minute channel will run up through the sensitive structures, infection penetrates and an inflammatory reaction is set up; this will be deep and difficult to resolve. When corns or bruising is present the horse or pony will usually flinch when the area is tapped with a hammer, and lameness will probably increase with work, especially on uneven ground.

Treatment is to remove the shoe and pare down the horn to reveal the red-tinged horn—evidence of blood seeping into the horn from the damaged blood vessels above. In the majority of cases with simple bruising a dry corn will result, and all that is required is to remove the shoe and pare down the discoloured part to relieve tension. Several days rest should then be given. When the corn is found to be 'moist', indicating suppuration and deep infection, then veterinary advice will be needed. Soaking the foot in a bucket of water to which antiseptic disinfectant has been added will be useful, and poulticing over a long period may be needed. Tetanus immunization should also be checked.

The vulnerability of the sole and underlying structures to damage and injury is largely dependent on the shape and character of the foot. Feet with a good, strong, concave dome to the sole are much better able to withstand shocks and blows than the flattened types of feet with weaker and thinner soles. For this reason, if a horse is to be used on rough, uneven going I dislike the wide-open type of foot favoured by many, as it is an invitation to injury. Shallow soles of this type are often given a measure of protection by a leather pad being fitted across the sole under the shoe to help prevent stones causing corns or bruising. However, as this invites dirt to build up underneath that is difficult to remove it can cause other problems. I personally favour a bell-shaped hoof, and have a particular preference for the Spanish Andalusian and related breeds that have very hard feet with high-vaulted soles. This type of foot does have somewhat higher heels and can look 'boxy', but it should not be confused with the 'blocky' feet that develop with navicular disease

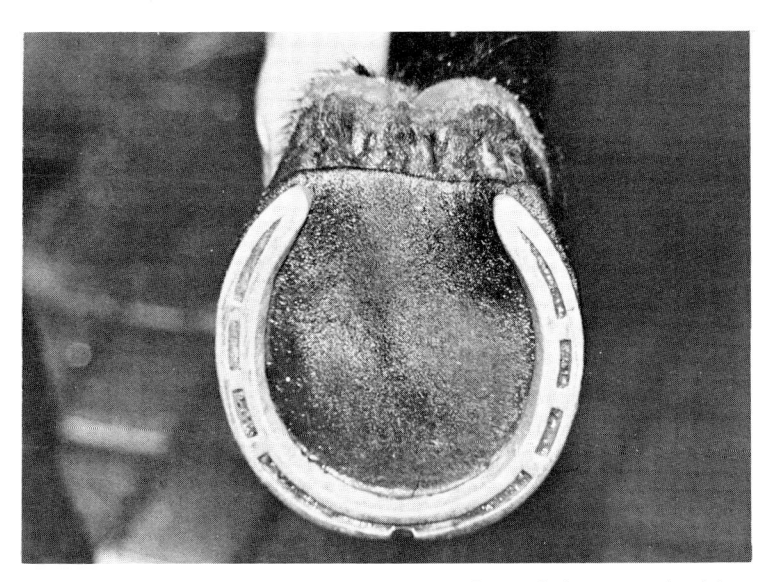

A leather pad fitted over the sole beneath the shoe to help prevent bruising.

in other horses; the incidence of navicular disease in horses with high-domed, bell-shaped hooves is lower than in the low-heeled, flat-footed types.

Thrush

When the fluid-secreting glands of the frog become irritated and impeded there is a build-up of fluid, followed by a chemical and bacteriological breakdown known as thrush. This gives rise to a sponginess of the frog, accompanied by an offensive smell. It can occur in any foot but is more common in the hind feet, probably because they are less 'open' and debris can thus become packed into them more easily. It is the presence in and around the frog of decomposing organic matter that accentuates the frog's deterioration and causes the foetid smell. Good management therefore plays an important part in preventing the disease, and a horse or pony's feet should be inspected and picked out twice a day in order to prevent a build-up of dung, stale urine, etc.

Once thrush has developed, the treatment is to clean out the foot thoroughly and pare off any loose or ragged pieces of the frog. Using a stiff scrubbing brush, give the infected area a thorough cleaning with water to which a strong antiseptic-disinfectant has been added. This treatment should be continued morning and night until the foot returns to a healthy condition, which will usually be within several days. It obviously helps if the animal is kept where it will not have to stand in dung and other decaying matter. Cod liver

A neglected foot, showing damage known as seedy toe.

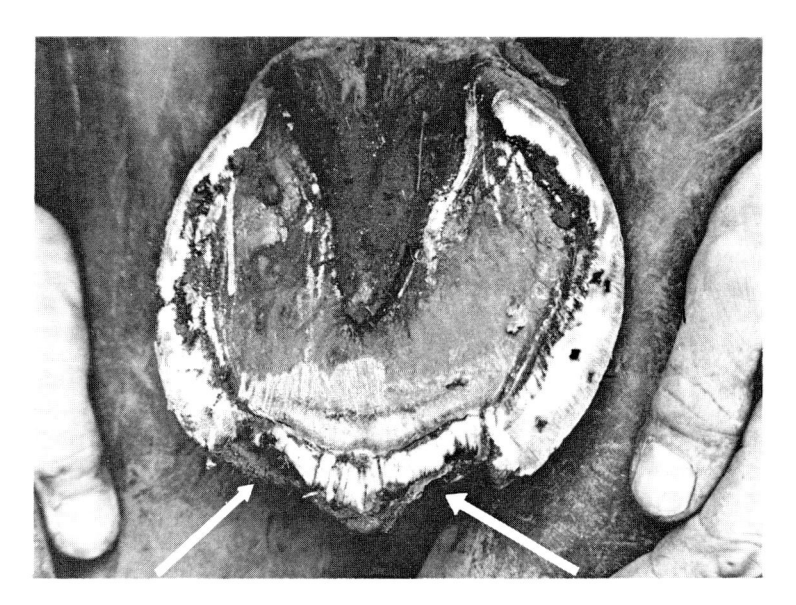

The same foot after it has been dressed by the farrier to enable a shoe to be fitted.

oil should be fed daily to help quicken the growth of the new frog and surrounding tissue, or a feed supplement extracted from seaweed specifically produced to promote healthy horn and similar tissue growth may be given.

Unless thrush has been present for a long time and is extremely bad a horse or pony will not become lame from this disease. Always remember the importance of keeping an active, healthy frog maintaining good pressure. This will go a long way towards preventing all kinds of lower leg troubles. Do not confuse the ragged appearance of a frog with small 'torn' pieces hanging from it as one having thrush. A normal, healthy frog will shed its outer layer in this way from time to time as the structure is renewed from within. Thrush cannot be mistaken because of its offensive smell.

Contracted heels

Lack of frog pressure, perhaps as a sequel to thrush, can contribute to contracted heels, which as the name suggests is a narrowing of the heels, usually those of the front feet. It results from the drying out of the horn through excessive rasping of its outer surface. Faulty shoeing that leaves the toes too long, thus losing the balance of expansion and contraction, can also contribute to contracted heels, as can the lost frog pressure resulting from setting up the heels too high. If any of these practices is responsible for the condition developing the remedy is obvious. Your blacksmith should be able to correct the abnormality, but beware of lowering the heel excessively in an attempt to bring back frog pressure: this could leave the heel so shallow that the hair line is almost at ground level, which could cause some of the other problems described. A shallow heel that has been lowered by cutting or rasping away the bars will be weakened and the sole will lose its domed shape. This will probably lead to injury of the foot with resulting lameness, whereas a contracted heel rarely does so. Much will depend on the natural shape of the foot, and any tendency to contracted heels should be discussed at length with an experienced blacksmith in order to bring about a gradual improvement.

Cracks and splits

The hoof wall will often suffer from cracks extending upwards from its lower permimeter; these are usually superficial except where the horny wall splits and peels back along the bottom edge. When a crack extends right through the wall it will cause the horny rim to break away, making shoeing difficult as the crack often appears where the shoe needs to be nailed on. There are several possible causes; working on rough stony going; shoeing that leaves the outer wall overhanging the shoe; dry, brittle horn which does not have enough tough elasticity. When the horn seems dry make sure that the outer layer of protective coating is not being rasped off when the horse or pony is shod. Also look to the animal's diet, and make sure that any vitamin deficiencies are made good by green feed, special supplements or cod liver oil. Hoof dressings will also be of value. My own method of hoof management includes a regular daily dressing of Stockholm tar, Newmarket grease and Cornucrescine mixed together in equal proportions to form a paste with enough cod liver oil added until it becomes workable. As I buy large tins of cod liver oil for general feed supplement purposes it is always to hand, and a small tin of each of the other ingredients is all that is needed to provide a constant supply of the mixture. I have used this for years and find it gives a pleasing appearance to the hoof, helps prevent brittleness and is much cheaper than the various proprietory brands of hoof oil.

Sandcrack

This term is used to describe a split in the hoof that occurs at the top of the horny wall and spreads downwards. In this location it is more serious than cracks or splits along the bottom edge of the wall. A true sandcrack may extend from top to bottom of the hoof, and may also be deep enough to involve the inner sensitive tissues, when foreign bodies such as dirt and grit can gain access and infection will

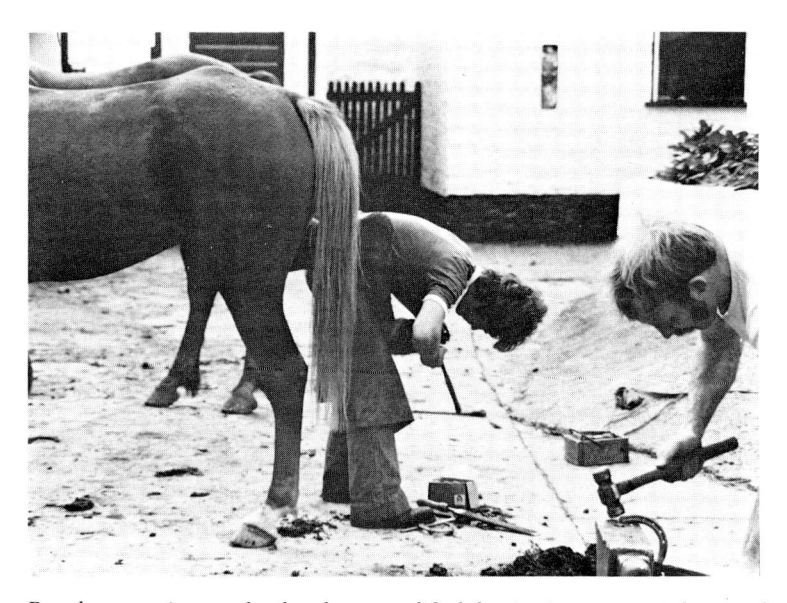

Regular attention to the feet by a qualified farrier is an essential part of keeping an animal sound.

result. The horse or pony will be in pain once this happens, and lameness will be present. A sandcrack is usually produced as the result of injury to the coronary band. It is here that the horny papillae or 'tubes' are produced that grow down to form the wall of the hoof. If there is injury to this area, perhaps resulting from a blow from the opposite foot, then the secretory function is interrupted. In serious cases where injury actually destroys part of the secretory band, a permanent fissure in the hoof wall will result as no more horn will grow down from that particular spot. Usually, however, the sandcrack is the result of a temporary arresting of the new papillae, which grow normally again when the injury heals. Painting the hoof with the dressing previously described is beneficial, and applying an antiseptic cream to the injured part of the coronary band will speed up healing. In cases of superficial sandcrack two grooves are sometimes cut in the horny wall on each side of the crack to form a 'V' at its lower limit. The object of this is to remove pressure from the edges of the crack and so isolate it and prevent it deepening. In severe sandcrack the two free extremities of the crack will have to be clipped together by the blacksmith. Sandcrack is an unsoundness; if buying a horse from a doubtful source all the feet should be closely examined to make sure that this condition has not been disguised by the crack being filled with pitch or wax and then painted over with hoof dressing.

Over-reaching, brushing and forging

There are a number of injuries sustained by the feet and lower leg that have various causes and are given a variety of names. Over-reaching occurs when a hind foot reaches too far forwards and strikes the hollow of the forefoot as it is raised from the ground. This usually occurs at the gallop, and the damage can sometimes be caused to the tendons at the back of the foreleg as high up as the knee. It can also be caused by a hind foot striking a front leg when landing after jumping. Horses with short backs and long legs, or with long hind legs in proportion to the front, are often said to be most prone to over-reaching. Special shoeing, setting the hind shoe well back and rounding the inside edge of the toe, will help to prevent injury when contact is made, as it is the inner edge of the toe of the hind shoe that does the damage to the front heel. Special boots made of rubber or similar material can also be worn by the horse to protect the front heels. When the injury is confined to the heel of the front foot it usually results in a skin wound, very often with a piece of loose skin left hanging from the lower edge of the tear. This should be removed as there is little chance of it growing back, and if it were to do so it would probably leave a permanent blemish in the form of proud skin. Clean the wound with salt water, and apply a dressing with a suitable ointment or a poultice if there is likelihood of infection needed to be drawn out. As healing takes place use just a dry dressing to keep it clean so that excessive healing tissue is not encouraged to grow. Lacerations to the bulbs of the heel can be slow to heal as the area is naturally rather poorly supplied

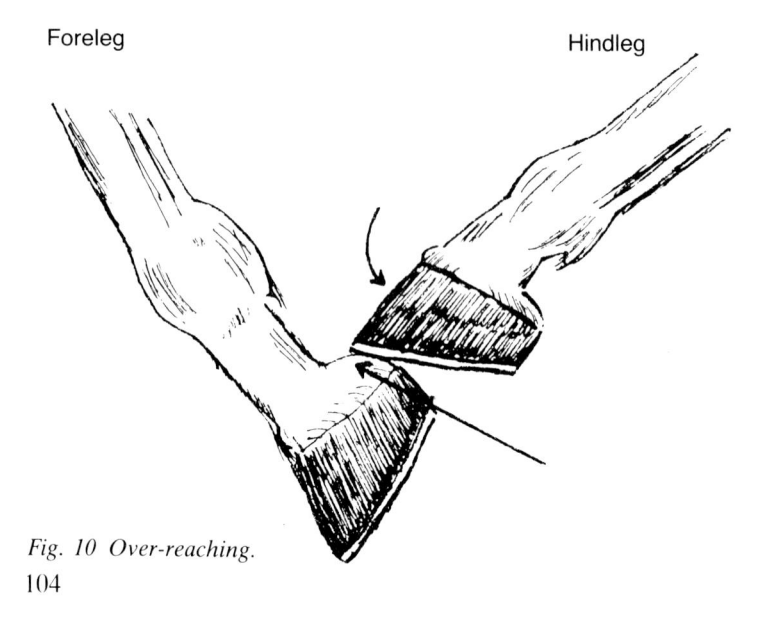

Fig. 10 Over-reaching.

with blood. Over-reach resulting in injury to the front fetlock or flexor tendons is much more serious. Although it may at first seem that only skin damage has resulted, actual damage beneath the skin can be extensive and there is the likelihood of damage to the synovial sheath; once the wound has been cleaned, therefore, it is best left open until a vet has been summoned.

Brushing

Brushing or, as it is sometimes called, 'cutting', is an injury to the inside of the fetlock when it is struck by the shoe of the opposite foot. It can happen with young horses or ponies when they are shod for the first time or when they are worked before maturity. Old horses can also suffer from this, as will unfit animals towards the end of a long and tiring period of work. Horses and ponies with a straight action will either grow out of it or it will cease to occur as they become properly fit. Animals with faulty conformation (i.e. turned-out toes, especially when this is accompanied by a narrow chest) may always be liable to brushing, and such animals are often equipped with protective brushing boots. There are several varieties of these available, usually made from leather.

Forging

Sometimes also referred to as 'clacking', this is not an injury but the noise made by the hind and front shoes striking together when a horse or pony is trotting. Young and green animals will often be heard to do it and it also occurs with tiredness. It wil cease when the horse or pony gains strength and comes up to the bit. Special shoeing techniques can help but are mostly unnecessary and the problem will usually cease to annoy by training. The practice is, however, not without danger as the blow may loosen or knock off the front shoe, or in rare instances the front and hind shoes may lock together, resulting in the animal being brought down.

Stumbling

Like forging, this is not an injury but can lead to one, including injury to the rider. Careless action can result in stumbling and so too can 'daisy-cutting' when a horse with a long, low action puts its toe to the ground first. The foot slides forwards a short distance on striking the ground and if it comes up against an obstacle, such as may be found in rough, stony going, the foot will tip over and the horse will be propelled forwards beyond its normal base of support. For this reason, when riding in such conditions or on rough, undulating or hilly country I prefer a riding horse with more upright action; though such an animal will not be as fast at the gallop it will be safer and more sure-footed at all paces.

Index

Acidity, 22, 27, 30
Anaemia, 16, 30
Appetite, 15
Arthritis, 32, 80
At grass, 13, 42, 93, 94
Azoturia, 27, 31, 32

Barn, building of a, 18
Bedding, 11, 48, 86, 88
'Big leg', 81–84
Blistering, 72, 73
Blisters, from tack pressure, 38
Bog spavin, 80, 81
Bone spavin, 80
Boredom, 10
Bran mash, 49, 53, 81
'Breaking out', 49
Bruised sole, 98, 99
Bruises, 34, 35

Capped elbow, 86, 88
Capped hock, 86, 88
Colic, 27–30
Coldness, signs of, 12
Colds, 48
Constipation, 28
Contracted heels, 101, 102
Convalescence, 11
Corns, 98, 99
Coughs, 48, 50–53
Cracked heels, 41, 42
Crib biting, 10
Curb, 84

Denerving, (Neurectomy), 95, 96
Dermatitis, 42, 43
Diarrhoea, 28
Dietary imbalances, 32, 93, 94
Diet, feedstuffs, 13, 14
Digestion, 27
Digestive disturbances, 28, 31, 51

Equine influenza, 52–54

Feedstuffs, manufactured, 14
Firing, line, 73, 74
Firing pin, 73, 74
Frog, 89, 90, 100–102
Glands, lymphatic, 81–84
Glands, swollen, 52
Glucose, 49
Grain, the storage of, 21
Grain, suppliers of, 17

Hay, the storage of, 17
Hobday, 57
Hock lameness, 78–81, 84

Inflamation, 36, 51, 80, 91, 96, 98
Inoculation, 47, 53

Lameness, determining, 59–65
Laminitis, 14, 27, 92–94
Larynx, 55–57
Lice, 42

Ligaments, 66, 68, 70, 84
Lumps, 34, 37–40, 45, 46
Lymphangitis, 81–84

Melanomas, 45, 46
Minerals, 22
Mouth, troubles of the, 24
Mud fever, 41, 42
Muscle damage, 32

Navicular disease, 94, 95
Nerve-blocking, 60, 61
Nursing the stabled horse, 12

Occult spavin, 80
Overfeeding, 14, 81, 93
Over-reaching, 104

Pedal ostitis, 91, 92
Phosphorous, 22
Pneumonia, 54
'Pointing', 63, 94
Poultices, the need for, 36, 40, 97

Rain scald, 43
Rasping of the teeth, 26, 27
Respiratory noises, 55–58
Respiratory system, 48, 50, 51
Ringbone, 96
Ringworm, 43
Roaring, 52

Saddle cloths, 37
Sandcrack, 102, 103
Sarcoids, 44, 45
Shivering, 54
Shoes, uneven wear of, 78
Sidebone, 96
Skin, diseases of, 43–46
Skin, functions of the, 36, 37, 49

Skin, irritations of the, 37–41, 42, 47
Sore backs, 37–40
Spavin, 63
Splints, 76–78
Sprains, tendon, 66–68, 70–74
Strangles, 52, 57
Straw, the storage of, 17
'Sweating up', 49
Sweet itch, 42
Swellings, the treatment of, 36
Synovial enlargements, 84–88

Teeth, troubles of the, 24–26
Temperature, 15, 49, 54
Tendons, bruised, 35
Tendons, sprains, 66–68, 70–74
Tendons, working of the, 65–67
Tendons, wounding of the, 36
Tetanus, 36, 47, 53
Thoroughpin, 81
Thrush, 100, 101

Ulcers, 81–83

Vermin, prevention of, 21
Vitamins, 22

Warble fly, 40
Warts, 43
Weaving, 10
Whistling, 52
Windgalls, 84, 85
Wind, broken, 50, 51
Wind, sucking, 10
Wind, thick, 57
Worms, 22, 28, 30
Wounds, 35, 38–41, 47, 96, 97, 104